Praise for *Denialism* by Mi[...]

"Michael Specter has written a lucid and insightful book about a very frightening and irrational phenomenon—the fear and superstition that threaten human science and progress. A superb and convincing work."

—Malcolm Gladwell, *New Yorker* staff writer and author of *Outliers*, *Blink*, and *The Tipping Point*

"Specter offers strong arguments in many cases against the public's antagonism to science. . . . The book opens a discussion about areas of science that are often neglected while more high-profile scientific controversies—global warming, stem cells, and human evolution—hog the stage."   —*USA Today*

"Specter is both provocative and thoughtful in his defense of science and rationality—though he certainly does not believe that scientists are infallible. His writing is engaging and his sources are credible, making this a significant addition to public discourse on the importance of discriminating between credible science and snake oil."   —*Publishers Weekly*

"Brims with passion and many interesting facts."
                    —*The New York Times Sunday Book Review*

"Written by a *New Yorker* staff journalist who specializes in public health, this work reports on the influence of activists, often celebrities, who reject mainstream medical science. Collecting their attitudes under the rubric of 'denialism,' Specter relentlessly reproves their arguments, identifies the fears behind denialism, and refracts as refutation the views of genuine scientists who research the specific health topic at hand."   —*Booklist*

"*Denialism* tells stories I know well, at least in outline. But Michael Specter very valuably gathers them under one roof and gives them a name. Specter describes the increasing public willingness to deny the hard-won facts of science in favor of myths and shoddy investigation. In the process, the denialists are enabling disease and poverty, denying the advances of science to those in need."
—David Baltimore, president emeritus of biology, California Institute of Technology

"We are bombarded with information and misinformation about the foods we eat, the medicines we take, the water we drink, the very air we breathe. Michael Specter shows us how to accurately assess the impact of science on these and other essential elements of our daily lives. Written in clear and accessible language, this uniquely valuable book explains an often confusing world."
—Jerome Groopman, M.D., Recanati Professor, Harvard Medical School, and author of *How Doctors Think*

"Serious, thoughtful, evidence-based conversations about our technological choices are crucial if we hope to navigate the future. This book adds to that conversation."   —*Winnipeg Free Press*

PENGUIN BOOKS

DENIALISM

Michael Specter writes about science, technology, and global public health for the *New Yorker*, where he has been a staff writer since 1998. Specter previously worked for the *New York Times* as a roving correspondent based in Rome, and before that as the *Times*'s Moscow bureau chief. He also served as national science reporter for the *Washington Post* as well as the New York bureau chief. He has twice received the Global Health Council's Excellence in Media Award, first for his 2001 article about AIDS, "India's Plague," and secondly for his 2004 article "The Devastation," about the health crisis in Russia and the country's drastic demographic decline. In 2002, Specter received the Science Journalism Award from the American Association for the Advancement of Science, for his article "Rethinking the Brain," about the scientific basis of how we learn. His work has been included in numerous editions of *The Best American Science Writing*.

DENIALISM

HOW IRRATIONAL THINKING

HARMS THE PLANET

AND THREATENS OUR LIVES

Michael Specter

PENGUIN BOOKS

PENGUIN BOOKS

Published by the Penguin Group

Penguin Group (USA) Inc., 375 Hudson Street, New York, New York 10014, U.S.A.
Penguin Group (Canada), 90 Eglinton Avenue East, Suite 700, Toronto,
Ontario, Canada M4P 2Y3 (a division of Pearson Penguin Canada Inc.)
Penguin Books Ltd, 80 Strand, London WC2R 0RL, England
Penguin Ireland, 25 St Stephen's Green, Dublin 2, Ireland (a division of Penguin Books Ltd)
Penguin Group (Australia), 250 Camberwell Road, Camberwell,
Victoria 3124, Australia (a division of Pearson Australia Group Pty Ltd)
Penguin Books India Pvt Ltd, 11 Community Centre,
Panchsheel Park, New Delhi – 110 017, India
Penguin Group (NZ), 67 Apollo Drive, Rosedale, North Shore 0632,
New Zealand (a division of Pearson New Zealand Ltd)
Penguin Books (South Africa) (Pty) Ltd, 24 Sturdee Avenue,
Rosebank, Johannesburg 2196, South Africa

Penguin Books Ltd, Registered Offices:
80 Strand, London WC2R 0RL, England

First published in the United States of America by The Penguin Press,
a member of Penguin Group (USA) Inc. 2009
This edition with a new afterword published in Penguin Books 2010

1 3 5 7 9 10 8 6 4 2

THE LIBRARY OF CONGRESS HAS CATALOGED THE HARDCOVER EDITION AS FOLLOWS:
Specter, Michael.
Denialism : how irrational thinking hinders scientific progress, harms the planet,
and threatens our lives / Michael Specter.
p. cm.
Includes bibliographical references and index.
ISBN 978-1-59420-230-8 (hc.)
ISBN 978-0-14-311831-2 (pbk.)
1. Science—Social aspects. 2. Research—Forecasting. 3. Belief and doubt.
4. Science—Philosophy. I. Title.
Q175.5.S697   2010
306.4'5—dc22   2009028489

Printed in the United States of America
DESIGNED BY AMANDA DEWEY

To Emma, who all by herself
is reason enough to reject denialism.

CONTENTS

DENIALISM

INTRODUCTION

Ten years ago, while walking through Harvard Yard, I saw a student wearing a button that said "Progressives against Scientism." I had no idea what that meant, so I asked him. Scientism, he explained, is the misguided belief that scientists can solve problems that nature can't. He reeled off a series of technologies that demonstrated the destructiveness of what he called the "scientific method approach" to life: genetically modified foods, dams, nuclear power plants, and pharmaceuticals all made the list. We talked for a few minutes, then I thanked him and walked away. I didn't understand how science might be responsible for the many scars humanity has inflicted upon the world, but students have odd intellectual infatuations, and I let it slip from my mind.

Over the next few years, while traveling in America and abroad,

I kept running into different versions of that student, people who were convinced that, largely in the name of science, we had trespassed on nature's ground. The issues varied, but not the underlying philosophy. Society had somehow forgotten what was authentic and there was only one effective antidote: embrace a simpler, more "natural" way of life. No phenomenon has illustrated those goals more clearly than persistent opposition to genetically engineered food. "This whole world view that genetically modified food is there so we have no choice but to use it is absolutely terrifying and it is wrong," Lord Peter Melchett, a former British Labour minister, told me when I met him a few years ago.

Today, Lord Melchett, whose great-grandfather founded one of the world's largest chemical companies, is policy director of the British Soil Association, the organic food and farming organization. The first time we spoke, however, he served as executive director of Greenpeace, where he was in the midst of leading a furious campaign against Monsanto (which he referred to as "Monsatan") to rid the world of genetically engineered foods. "There is a fundamental question here," he said. "Is progress really just about marching forward? We say no. We say it is time to stop assuming that discoveries only move us forward. The war against nature has to end. And we are going to stop it."

I felt then—as I do now—that he had gotten it exactly wrong; scientists weren't waging a war at all, he was—against science itself. Still, I saw Lord Melchett as a quaint aristocrat who found an interesting way to shrug off his family's industrial heritage. His words were hard to forget, though, and I eventually came to realize why: by speaking about a "war against nature," he had adopted a system of belief that can only be called denialism. Denialists like

Lord Melchett replace the rigorous and open-minded skepticism of science with the inflexible certainty of ideological commitment.

We have all been in denial at some point in our lives; faced with truths too painful to accept, rejection often seems the only way to cope. Under those circumstances, facts, no matter how detailed or irrefutable, rarely make a difference. Denialism is denial writ large—when an entire segment of society, often struggling with the trauma of change, turns away from reality in favor of a more comfortable lie.

Denialism comes in many forms, and they often overlap. Denialists draw direct relationships where none exist—between childhood vaccinations, for example, and the rising incidence of diseases like diabetes, asthma, and autism. They conflate similar but distinct issues and treat them as one—blending the results of different medical studies on the same topic, or confusing a general lack of trust in pharmaceutical companies with opposition to the drugs they manufacture and even to the very idea of science.

Unless data fits neatly into an already formed theory, a denialist doesn't really see it as data at all. That enables him to dismiss even the most compelling evidence as just another point of view. Instead, denialists invoke logical fallacies to buttress unshakable beliefs, which is why, for example, crops created through the use of biotechnology are "frankenfoods" and therefore unlike anything in nature. "Frankenfoods" is an evocative term, and so is "genetically modified food," but the distinctions they seek to draw are meaningless. All the food we eat, every grain of rice and ear of corn, has been manipulated by man; there is no such thing as food that *hasn't* been genetically modified.

Our ability to cut genes from one organism and paste them into another has transformed agriculture. But it is a change of degree, not of type. Denialists refuse to acknowledge that distinction, in part because it's so much simpler to fix blame on a company, an institution, or an idea than to grapple with a more complicated truth: that while scientific progress has brought humanity immense wealth and knowledge, it has also caused global pollution severe enough to threaten the planet. Denialists shun nuance and fear complexity, so instead of asking how science might help resolve our problems, they reject novel strategies even when those strategies are supported by impressive data and powerful consensus.

Until I learned about Holocaust deniers, it never occurred to me that a large group could remain willfully ignorant of the most hideous truths. Then, twenty-five years ago, I began to write about people who refused to acknowledge that the human immunodeficiency virus caused AIDS, despite what, even then, was an overwhelming accretion of evidence. Holocaust deniers and AIDS denialists are intensely destructive—even homicidal—but they don't represent conventional thought and they never will. This new kind of denialism is less sinister but more pervasive than that.

My unusual encounter at Harvard came back to me a few years ago, and I started to think about writing this book. I kept putting it off, though. Some of the delay was due simply to procrastination. But there was another, more important reason for my hesitation: I had assumed these nagging glimpses of irrationality were aberrations, tiny pockets of doubt. Authority may be flawed, and science often fails to fulfill its promises. Nonetheless, I was convinced that people would come around to realizing that the "scientific method approach"—the disciplined and dispassionate search for knowledge—has been the crowning intellectual achievement of humanity. I guess I was in my own kind of denial, because even as things got worse I kept assuring myself that reason would prevail and a book like this would not be necessary.

Finally, a couple of years ago, I was invited to dinner at the home of a prominent, well-read, and worldly woman. She asked what I was working on and I told her that I had become mystified by the fact that so many Americans seemed to question the fundamental truths of science and their value to society. I mentioned as examples anxiety about agricultural biotechnology, opposition to vaccinations, and the growing power of the alternative health movement.

She suddenly became animated. "It's about time somebody writes the truth about these pharmaceutical companies," she cried. "They are evil, making vast sums from lifestyle drugs like Viagra and letting millions die instead of helping them. The government is no better; they are destroying our food supply and poisoning our water." Some years earlier she had been seriously ill, and she explained how she recovered: by taking dozens of vitamins every day, a practice she has never abandoned. With this woman's blessing, her daughter, who had just given birth, declined to vaccinate her baby.

The woman didn't actually say, "It's all a conspiracy," but she didn't have to. Denialism couldn't exist without the common belief that scientists are linked, often with the government, in an intricate web of lies. When evidence becomes too powerful to challenge, collusion provides a perfect explanation. ("What reason could the government have for approving genetically modified foods," a former leader of the Sierra Club once asked me, "other than to guarantee profits for Monsanto?")

"You have a point," I told the woman. "I really ought to write a book." I decided to focus on issues like food, vaccinations, and our politically correct approach to medicine, because in each of those arenas irrational thought and frank denialism have taken firm root. Today, anyone who defends science—particularly if he suggests that pharmaceutical companies or giant agricultural conglomerates may not be wholly evil—will be called a shill.

That's denialism, too—joined as it often is with an almost religious certainty that there is a better, more "natural" way to solve our medical and environmental problems. Answers are rarely that simple, though. Even in the case of the drug Vioxx, which I describe in the next chapter—where Merck was as guilty of malfeasance as a company can be—it's likely that had the drug remained on the market, it would have been responsible for a hundred times more good than harm.

The most blatant forms of denialism are rarely malevolent; they combine decency, a fear of change, and the misguided desire to do good—for our health, our families, and the world. That is why so many physicians dismiss the idea that a patient's race can, and often should, be used as a tool for better diagnoses and treatment. Similar motivations—in other words, wishful thinking—have helped drive the growing national obsession with organic food. We want our food to taste good, but also to be safe and healthy. That's natural. Food is more than a meal, it's about history, culture, and a common set of rituals. We put food in the mouths of our children; it is the glue that unites families and communities. And because we don't see our food until we eat it, any fear attached to it takes on greater resonance.

The corrosive implications of this obsession barely register in

America or Europe, where calories are cheap and food is plentiful. But in Africa, where arable land is scarce, science offers the only hope of providing a solution to the growing problem of hunger. To suggest that organic vegetables, which cost far more than conventional produce, can feed billions of people in parts of the world without roads or proper irrigation may be a fantasy based on the finest intentions. But it is a cruel fantasy nonetheless.

Denialist arguments are often bolstered by accurate information taken wildly out of context, wielded selectively, and supported by fake experts who often don't seem fake at all. If vast factory farms inject hormones and antibiotics into animals, which is often true and always deplorable, then *all* industrial farming destroys the earth and *all* organic food helps sustain it. If a pricey drug like Nexium, the blockbuster "purple pill" sold so successfully to treat acid reflux disease, offers few additional benefits to justify its staggering cost, then all pharmaceutical companies always gouge their customers and "natural" alternatives—largely unregulated and rarely tested with rigor—offer the only acceptable solution.

We no longer trust authorities, in part because we used to trust them too much. Fortunately, they are easily replaced with experts of our own. All it takes is an Internet connection. Anyone can seem impressive with a good Web site and some decent graphics. Type the word "vaccination" into Google and one of the first of the fifteen million or so listings that pops up, after the Centers for Disease Control, is the National Vaccine Information Center, an organization that, based on its name, certainly sounds like a federal agency. Actually, it's just the opposite: the NVIC is the most powerful anti-vaccine organization in America, and its relationship with the U.S. government consists almost entirely of opposing federal efforts aimed at vaccinating children.

IN 2008, America elected a president who supports technological progress and scientific research as fully as anyone who has held the office. Barack Obama even stressed science in his inaugural address. "We will restore science to its rightful place and wield technology's wonders to raise health care's quality . . . and lower its costs," he said. "We will harness the sun and the winds and the soil to fuel our cars and run our factories. And we will transform our schools and colleges and universities to meet the demands of a new age."

Obama realizes the urgency with which we need to develop new sources of energy. That is why he frequently compares that effort to America's most thrilling technological achievement: landing men on the moon. Obama has assembled a uniformly gifted team of scientific leaders, and when he speaks publicly about issues like swine flu or HIV, the president routinely makes a point of saying that he will be guided by their advice.

That is quite a departure from the attitude of his predecessor, who, in one of his first major initiatives, announced that he would prohibit federal funding for research on new stem cell lines. George W. Bush encouraged schools to teach "intelligent design" as an alternative to the theory of evolution, and he all but ignored the destruction of our physical world. His most remarkable act of denialism, however, was to devote one-third of federal HIV-prevention funds to "abstinence until marriage" programs.

The Bush administration spent more than $1 billion on abstinence-only programs, despite data from numerous studies showing that they rarely, if ever, accomplish their goals. Nevertheless, during the Bush administration, family planning organizations in the developing world were denied U.S. grants if they so much as discussed abortion with their clients. President Obama began at once to reverse that legacy and restore the faith in progress so many people had lost.

I wish I could say that he has helped turn back the greater tide of denialism as well. That would be asking too much. Despite the recession, sales of organic products have continued to grow, propelled by millions who mistakenly think they are doing their part to protect their health and improve the planet. Supplements and vitamins have never been more popular even though a growing stack of evidence suggests that they are almost entirely worthless. Anti-vaccination conspiracy theorists, led by the tireless Jenny McCarthy, continue to flourish.

So does denialism, abetted by some of the world's most prominent celebrities. Oprah Winfrey, for one, has often provided a forum for McCarthy on her show, but she intends to do more: in early 2009, Winfrey's production company announced that it had hired McCarthy to host a syndicated talk show and write a blog, providing two new platforms from which she can preach her message of scientific illiteracy and fear.

This antipathy toward the ideas of progress and scientific discovery represents a fundamental shift in the way we approach the world in the twenty-first century. More than at any time since Francis Bacon invented what we have come to regard as the scientific method (and Galileo began to put it to use), Americans fear science at least as fully as we embrace it. It is a sentiment that has turned our electrifying age of biological adventure into one of doubt and denial. There have always been people who are afraid of the future, of course—Luddites, ignorant of the possibilities of life on this planet and determined to remain that way. No amount of data will convince climate denialists that humans have caused the rapid, devastating warming of the earth. And no feat of molecular genetics will make a creationist understand that our species has evolved over billions of years, along with every other creature.

Common strains of denialism are even more troubling, though, because they show what happens when unfettered scientific achievement bumps up against the limits of human imagination. Manipulating the genes of cows or corn was only a first step. Today, we routinely intrude on every aspect of human and natural life. That fact traumatizes people—and not entirely without reason. Mary Shelley couldn't have imagined what goes on in thousands of laboratories today. Scientists all over the world are resurrecting viruses that have been extinct for millions of years. They are constructing organs out of spare parts, and it is only a matter of time (and not much time either) before synthetic biologists design, then grow, entirely new forms of life—organisms that have never before existed in the natural world. The speed at which all this is happening has made many people fear that we are about to lose control, not only over the world we have always viewed as our dominion, but of human life as well.

Nothing scares us quite as much. Controlling life is something we have attempted since we domesticated cattle and began to grow food. The scientific revolution helped solidify the idea that our species was in command and, as Bacon put it in *The New Atlantis*, able to "establish dominion over nature and effect all things possible." Yet there have always been committed efforts at stopping the march of technology.

In 1589, Queen Elizabeth refused to fund a project to make a knitting machine, saying, "My lord, I have too much love for my poor people who obtain their bread by knitting to give money that will forward an invention which will tend to their ruin by depriving them of employment." Three centuries later, in 1863, Samuel Butler became the first to write about the possibility that machines might evolve through Darwinian selection. Although many readers thought he was joking in his essay "Darwin Among the Machines," Butler was serious (and astonishingly prescient):

> There are few things of which the present generation is more justly proud than of the wonderful improvements which are daily taking place in all sorts of mechanical appliances. . . . Day by day, however, the machines are gaining ground upon us; day by day we are becoming more subservient to them; more men are daily bound down as slaves to tend them, more men are daily devoting the energies of their whole lives to the development of mechanical life. The upshot is simply a question of time, but that the time will come when the machines will hold the real supremacy over the world and its inhabitants is what no person of a truly philosophic mind can for a moment question.

If anything, that fear is more pronounced today (and more understandable) than ever before. Denialism is often a natural response to this loss of control, an attempt to scale the world to dimensions we can comprehend. Denialism is not green or religious or anti-intellectual, nor is it confined to utopian dreamers, agrarians, or hippies. It is not right- or left-wing; it is a fear expressed as frequently and with as much fervor by Oxford dons as by bus drivers.

The fear has seeped across Britain, Europe, and the developing world. But nowhere is it more evident than in the United States, a country that has always defined itself by its notion of progress and technological prowess. We may be a nation of immigrants, but more than that we are the nation that invents: from refrigerators to resistors, antibiotics, jets, and cell phones, to the computer software that governs much of our lives and the genetic sequencing technology that will soon begin to do so. What would have seemed like sorcery a century ago is now regarded simply as fact. In 1961, Arthur C. Clarke famously wrote that "any sufficiently advanced technology is indistinguishable from magic."

Who could make such a statement today? What would magic look like to us? It has become routine to deliver babies months before they are considered alive—not to mention to keep people breathing long after they are, in any meaningful sense, dead. My grandfather died in 1962 at the age of sixty-six. That was exactly how long men born at the turn of the twentieth century were "expected" to live, and while he was mourned, nobody considered his death premature. They certainly would have today, though. Just forty-six years later, a healthy fifty-year-old man can expect to live to the age of eighty.

At least since the Enlightenment, when science effectively replaced religion as the dominant ideology of mankind, progress has been our purpose. We have moved from the discovery of the compass (and our sense of where we are in the physical world) to the invention of gunpowder, to the astonishing ability to take pictures that see through human flesh—only to arrive at the defining event of the twentieth century: the splitting of the atom. As Manhattan Project scientists gathered in New Mexico on July 16, 1945, to await results from the first test of the atomic bomb, they were anxious and afraid. Many took bets on whether they were about to set the sky ablaze and destroy the world.

New technologies are always accompanied by new risks and at least one deeply unsettling fact: once you invent something you cannot uninvent it. That sounds simple, but since that day in New Mexico more than half a century ago that knowledge has changed society, planting seeds of fear into even the most promising discoveries. The superpowers may have averted a cold war and dismantled many of the nuclear weapons that had threatened to annihilate us. But they didn't uninvent them and they never could. H. G. Wells said that civilization is a race between education and catastrophe. He was right. Even more than that, though, civilization is a race between innovation and catastrophe.

That race only grows more frantic. Global nuclear war, while by no means impossible, is a less likely prospect than it was twenty years ago. But there is nothing unlikely about assembling life from scratch, cloning copies of ourselves, or breeding extinct animals. In November 2007, for the first time, researchers successfully cloned embryos from the single cell of an adult monkey. The work put an end to any debate about whether primates—the group that includes not only monkeys but men—are biologically capable of being turned into clones. Faust and Frankenstein have been with us for a long time. But the wall between science fiction and reality has practically vanished, and there is evidence of that in even the most trivial places.

The 2007 film *I Am Legend* was hardly a cinematic master-piece, but it opens with a scene in which a doctor explains to a TV news anchor how she was able to cure cancer by mutating the measles virus and harnessing its destructive power. She tells him that measles is like "a fast car with a madman at the wheel," but her team believed it could be used for good if "a cop were driving it instead." So they used the virus to cure cancer, which was wonderful until the misprogrammed organism wiped out nearly everyone on earth. Like Godard's film *Le Nouveau Monde* and the original 1954 book on which both films were based, *I Am Legend* is pure fiction, another story of a virus gone wild. That doesn't mean it will be fiction tomorrow.

In early 2009, a team from the Mayo Clinic reported that certain measles strains could prove effective as a treatment for cancer. "These viral strains could represent excellent candidates for clinical testing against advanced prostate cancer," said Evan-thia Galanis, the senior author of the paper. The viral strains were inactivated, harmless, and well contained in a highly secure lab. Nonetheless, it is hard not to recoil when life imitates art so faithfully.

"Of course this is all possible," Drew Endy said when I asked him whether the theoretical threats posed by the new science of synthetic biology were real. "If we don't want to exist, we can stop existing now." Endy is a biological engineer at Stanford University who is essentially attempting to turn human cells into software that we can program. Instead of producing iTunes or spreadsheets, however, this software would attack tumors, repair arteries clogged with cholesterol, and prevent diseased cells from destroying the immune system. Endy is an optimist, but he readily acknowledges the dangers associated with his work. "Why wouldn't we be afraid?" he said. "We are speaking about creating entirely new forms of life."

FIFTY YEARS AGO, we venerated technology. At least until we placed our feet on lunar soil, our culture was largely one of uncritical reverence for the glories that science would soon deliver. The dominant image of popular American culture was progress. TV shows like *Star Trek* and *The Jetsons* were based on a kind of utopian view of the scientific future. Even the Flintstones were described as a "modern" Stone Age family. We were entering an era without disease or hunger. If we ran out of water we would siphon salt from the seas and make more; if nature was broken we could fix it. If not, we could always move to another planet.

That vision no longer seems quite so enchanting. No doubt our expectations were unreasonable—for science and for ourselves. We also began to recognize the unintended consequences of our undeniable success. About a month before Neil Armstrong made his large step on the moon, the heavily polluted Cuyahoga River erupted in flames near Cleveland, creating an indelible image of industry at war with nature. A few years later, in 1976, Karen Ann Quinlan was removed from life support, igniting the first horrific battle of the modern era over how we live and die. The end of the decade was marked by the ghastly accident at Three Mile Island, which showed more clearly than ever that the effects of the Industrial Revolution were not all benign. The thalidomide disaster, mad cow disease, even the dramatic and sustained lies of Big Tobacco have all contributed to the sense that if the promise of science wasn't a lie, it wasn't exactly the truth either.

Today the image of a madman whipping up a batch of smallpox, or manufacturing an effective version of bird flu in his kitchen, while not exactly as easy as baking a cake, is no longer so far-fetched. Indeed, if there is anything more frightening than the threat of global nuclear war, it is the certainty that humans not only stand on the verge of producing new life forms but may soon be able to tinker with them as if they were vintage convertibles or bonsai trees.

Our technical and scientific capabilities have brought the world to a turning point, one in which accomplishments clash with expectations. The result often manifests itself as a kind of cultural schizophrenia. We expect miracles, but have little faith in those capable of producing them. Famine remains a serious blight on humanity, yet the leaders of more than one African nation, urged on by rich Europeans who have never missed a meal, have decided it would be better to let their citizens starve than to import genetically modified grains that could feed them.

Food is a compelling example of how fear has trumped science, but it is not the only evidence that we are waging a war against progress, rather than, as Peter Melchett would have it, against nature. The issues may be complex but the choices are not: we are either going to embrace new technologies, along with their limitations and threats, or slink into an era of magical thinking. Humanity has nearly suffocated the globe with carbon dioxide, yet nuclear power plants that produce no such emissions are so mired in objections and obstruction that, despite renewed interest on every continent, it is unlikely another will be built in the United States. Such is the opposition to any research involving experiments with animals that in scores of the best universities in the world, laboratories are anonymous, unmarked, and surrounded by platoons of security guards.

For hundreds of years we had a simple but stunningly effective approach to our interaction with the physical world: what can be understood, and reliably repeated by experiment, is what nature regarded as true. Now, at the time of mankind's greatest scientific advances (and our greatest need for them), that deal is off. Snake oil salesmen may be old news in America, but today quacks—whose research is even funded by the federal government—take out ads in the *New York Times* denouncing scientists who rely on evidence-based medicine to treat our most devastating diseases.

We are now able to stare so deeply into the molecular history of the human genome that, peering one hundred million years into the past, we can see that we shared a common ancestor with the elephant. Scientists are tantalizingly close to understanding how the trillions of cells in our bodies work and interact with each other. Nonetheless, in 2007, a $27 million Creation Museum opened in Kentucky, complete with bumper stickers that proclaim "We're Taking Dinosaurs Back." That's fitting, since no matter how you ask the question, at least one in three American adults rejects the concept of evolution, believing instead that humans descended from heaven several thousand years ago in our present form.

Science and religion have always clashed and always will. Einstein put it best: science without religion is lame, religion without science is blind. In the past, that conflict, while often painful, never managed to derail progress. We can no longer say that. If anything, our increasingly minute knowledge of the origins of humanity has served only to fuel the intelligent design movement, not to dampen it. In 2005, when the American Museum of Natural History mounted the most significant exhibition ever devoted to Charles Darwin, the leadership there couldn't find a single corporate sponsorship, as they always had been able to do in the past. Few American companies were willing to risk a boycott staged by those who object to the theory of evolution.

Denialism must be defeated. There is simply too much at stake to accept any other outcome. Who doesn't have a family member with diabetes, Parkinson's disease, Alzheimer's, or some form of cancer? When faced with genuine solutions (not just promises) to such terrible fates, few will continue to question the value of stem cell research or cloning. Even Nancy Reagan, whose husband served as commander-in-chief of the American war against legal abortion, became an ardent and vocal supporter of stem cell research after watching him submit to the dark fog of Alzheimer's disease.

We have acquired more knowledge in the past decade than in the previous two centuries. Even bad news soon proves its worth. Look at avian influenza: bird flu may cause a devastating epidemic. Viruses, like earthquakes and volcanic eruptions, will always be part of life on earth. (Not long before he died, Nobel Prize–winning biologist Joshua Lederberg told me that the "single biggest threat to man's continued dominance on this planet is the virus." He was not alone in believing that.) Nonetheless, avian influenza is the first potential pandemic in the history of humanity that can be understood even before it becomes contagious. Researchers have mapped every gene and protein in the virus and are well on their way to developing a vaccine.

Science has slowly come to define us. In 1959, C. P. Snow delivered his "Two Cultures" speech at Cambridge University, in which he asserted that the chasm between the worlds of science and the humanities was making it hard to solve the earth's most pressing problems. He had a point at the time. But we don't have two cultures anymore, we have one. Students in many classrooms seek answers to the remaining intricacies contained within the human genome, and if they don't understand their research they can always turn to the Internet to find an eager tutor from one of a dozen nations. In India and China, young engineers and biologists use Skype to conduct videoconferences with colleagues from Boston to Berlin. It costs nothing. Two generations ago, in the unlikely event that their grandparents had known how to write a letter, they would never have been able to afford postage stamps or find a place to mail it.

Ultimately, dramatic achievements have always taken us past our fears and overcome denialism—because progress offers hope and for humans nothing beats hope. Fear might threaten progress; in the end, though, it won't prevent it. Not long ago, after publishing a piece in the *New Yorker* on scientists who were reconstructing extinct viruses, I received this letter: "Not discounting the great advances we have made with molecular biology over the last twenty-five years, I dare say the question remains . . . will this generation of scientists be labeled the great minds of the amazing genetic-engineering era, or the most irresponsible scientists in the history of the world? With the present posture of the scientific community, my money, unfortunately, is on the latter."

Those words might as well have been torn from a denialist instruction manual: change is dangerous; authorities are not to be trusted; the present "posture" of the scientific community has to be one of collusion and conspiracy. Most important, the facts are inside out, because "discounting" the great advances of molecular biology is exactly what the author of that letter did.

Scientists should do a better job of explaining the nature and the potential impact of their work (and so should those of us who write about science). We need to have open debates—on national television and guided by people like President Obama—about how to engage the future and make sense of both the possibilities and risks that lie ahead. Education will have to improve as well. But to call the group that has decoded the language of life, and has already begun to use that information to treat and prevent scores of diseases, "the most irresponsible scientists in the history of the world" is specious. Without the tools of molecular biology, we wouldn't have a clue how the AIDS virus works. Instead of having killed twenty-five million people in the twentieth century and infected an even larger number, the toll of such an unimpeded epidemic would have already numbered in the billions.

No achievement of modern technology, not even nuclear power, has been more bitterly disputed than our ability to alter the genetic composition of food or to create artificial products from human cells. Yet no discovery is more likely to provide solutions to the greatest threat the earth has ever faced: the rapid pace of global warming. If we do not develop clean technologies soon, our species won't survive.

And we are doing just that. Researchers throughout the country and the world are fabricating synthetic molecules that mimic those found in nature. Ten years ago there wasn't even enough basic knowledge to attempt this kind of research. These companies are not interested in making fake corn or pursuing effective ways to manufacture fertilizer for farmers who really don't need it. Instead, they seek to fuel cars and power factories without releasing greenhouse gases. That would not only keep us from blanketing the earth in carbon dioxide, but would go a long way toward counteracting the foolish and expedient decision to make fuel from the corn, sugar, and soybeans that the rapidly growing population of the world so desperately needs to eat.

To accomplish any of this we will have to recognize denialism when we see it. As a society and as individuals, that means asking tough, skeptical questions, then demanding answers supported by compelling evidence. When the government, a company, or any other group makes a claim, we need to scrutinize that claim with care but without passion. Most importantly, we must learn to accept data that has been properly judged and verified—no matter what it says, or how much we might have wished that it pointed in another direction.

I wonder, as the ice sheet in Greenland disappears, the seas rise, and our sense of planetary foreboding grows, will denialists consider the genetically engineered organisms that propel our cars and sustain our factories as a continuation of what Lord Melchett described as a war against nature? Or will they see them for what they are, the latest—and grandest—stage in our march toward human enrichment?

1

Vioxx and the Fear
of Science

The daily work of science can be repetitive and dreary. Even the most talented researchers spend the bulk of their lives bathed in the fluorescent light of the laboratory, hovering over a bench, staring at slides, and hunting for meaningful patterns in strings of numbers. Still, like many of his colleagues, the cardiologist Eric Topol had always dreamed that one day he might have his "eureka" moment—a flash of insight that would permit him to see clearly what others couldn't see at all. In 2001, Topol got his wish—but not in a way he had ever imagined; there were no shouts of joy, no elation or champagne, nothing of the kind. "I was just sad," he said, remembering the moment when he realized that one of the nation's most popular new medications was killing people. "Then I was angry, and eventually I became outraged."

At the time, Topol was chairman of the Cleveland Clinic's cardiology department, which he more than any other physician had transformed into one of the finest in American medicine. His research into how to prevent and treat heart attacks was highly valued and constantly cited. As perhaps the clinic's most visible face, Topol was already prominent. But it was his role in helping to expose the grave risks posed by the anti-inflammatory drug Vioxx that turned him into one of the country's best-known doctors. It also made him one of the most controversial, in part because he repeatedly stressed how little regard the Food and Drug Administration seemed to have for the hundreds of millions of people it had been created to protect.

The cloak of invincibility had long before been stripped from any government agency, replaced by the constants of doubt and denial; politicians, scientists, doctors, and lawyers are held in lower esteem today than at any time in decades. Yet no previous incident—not the explosion of the space shuttle *Challenger*, Ford's willingness to dump a death trap called the Pinto on the American public, not even the nuclear accident at Three Mile Island—demonstrates more vividly why that mistrust has become so pervasive.

Vioxx was introduced by Merck with great enthusiasm in 1999, one of a new class of drugs called cox-2 inhibitors, which were designed to interfere with an enzyme called cyclooxygenase-2, which, among more beneficial duties, produces chemicals that cause inflammation (and pain). Before Vioxx appeared, hundreds of thousands of people who suffered the debilitating effects of arthritis and other chronic ailments faced an unpleasant choice every day: they could take drugs like aspirin or Advil, or they could endure agony in order to avoid the bleeding ulcers and other serious stomach complications those drugs can cause. Vioxx was referred to as "super aspirin," which didn't seem like much of an exaggeration: in early studies it offered better pain relief than any traditional remedy, and was far less likely to disturb the stomach. The drug quickly came to be seen by those who needed it most as a kind of magic potion, one that only the tools of modern medicine could have produced. Driven in part by one of the most aggressive advertising campaigns in medical history, more than twenty million Americans took Vioxx at one time or another. In 2003 alone, Merck sold more than $2.5 billion worth of the drug.

Topol, who suffered from an arthritic knee, loved Vioxx. Even now he readily attests to its effectiveness. "Nothing worked as well for me before or since," he said. "Vioxx truly dulled the pain." One February morning in 2001, though, he noticed a report that struck him as odd. Topol had been invited to deliver a lecture about the future of cardiac care to a gathering at the Medical College of Georgia in Augusta. Over breakfast at his hotel, he started to page through the copy of *USA Today* that he found on his doorstep. One particular story leapt out at him. "It was about Vioxx," he said, "and this study," called VIGOR—Vioxx Gastrointestinal Outcomes Research—"which was intended to determine whether Vioxx really was easier on the stomach than other, less powerful nonsteroidal anti-inflammatory medication."

Between January and July 1999, researchers had followed eight thousand patients with rheumatoid arthritis. Half took Vioxx to control their pain; the other half took naproxen, which is sold over the counter as Aleve. (It was a large and fairly conventional randomly assigned, double-blind study—which meant that the patients had no idea which of the two drugs they were taking, and neither did their doctors.) The first time the safety committee assigned to monitor the study looked at the data, it found exactly what one might have expected: people in the Vioxx group were less likely to experience significant stomach distress than those who took Aleve.

The trial also showed something that had not been anticipated, and the news there was disturbing: participants who already suffered from heart disease were far more likely to have heart attacks if taking Vioxx than if they had been given Aleve. Nobody was sure why, and because Merck had never expressed concern about the drug's effect on the heart, there were no cardiologists on the safety committee (which was not unusual since that wasn't the purpose of the trial). Scientists wondered if the difference might have been due to the fact that people in the trial were required to stop taking aspirin, since it can lower the risk of heart attack or stroke. It was also possible that something previously unrecognized about the chemical composition of Aleve itself helped protect the cardiovascular system. (That would have provided a benign explanation for the differing rate of heart attacks, and Merck endorsed the hypothesis with great enthusiasm.)

"I was not a drug safety expert and I never even had any interest in the issue," Topol said. "My principal research was in heart disease and heart attacks, and that dates back more than twenty years." Topol had made a name for himself as a postdoctoral researcher at the University of California at San Francisco. He then took a job at Johns Hopkins University, where he became the first physician to treat heart attacks with the powerful clot-busting agent known as tPA; he also directed a pivotal study that compared the efficacy of that drug with an older treatment, streptokinase, in saving lives. In 1991, Topol moved to the Cleveland Clinic, where for the next fifteen years he served as chairman of the department of cardiovascular medicine.

What Topol saw in *USA Today* that morning in Augusta made no sense to him. "Why would a new anti-inflammatory agent prove less protective against heart attacks than the one you can buy at a pharmacy with no prescription?" he wondered. Yet, the newspaper report suggested that patients taking Vioxx were more than twice as likely to have heart attacks as those taking Aleve. For people with a history of heart disease, the risk was far higher. (Risk numbers don't mean much unless they are accompanied by some assessment of the statistical probability that those risks could occur by chance. In this study, which at the time was the largest ever conducted using Vioxx, that number was .002—two in one thousand. In other words, if the study were repeated a thousand times, results like those would appear by chance twice.) "I thought, that is interesting, they are saying a highly touted experimental drug was not as good as the one you buy in the drugstore," Topol said. "They weren't saying anything about Vioxx *causing* heart attacks, just that Aleve seemed better at preventing them."

The distinction was crucial because after the data was made public, Merck asserted, as it would for the following three years, that Vioxx posed no increased risk of heart attack or stroke. "It did seem strange," Topol said, "but I didn't give it a lot of thought. After all, that's why you do clinical studies—and so it looked like maybe Aleve had some protective effect we didn't know about. Surprising as that would have been, it certainly wasn't beyond the realm of possibility. Still, it wasn't my thing and I didn't dwell on it."

Topol delivered his address and returned to Cleveland, where Debabrata Mukherjee—"one of my fellows and a brilliant scientist"—had also seen the report which showed that people using Vioxx were far more likely to suffer from heart attacks than those taking over-the-counter pain medication. Mukherjee became intensely curious about the reasons for such surprising results. He dove into the data that Merck had been required to provide the FDA, and soon realized that the initial report failed to include all the essential information at Merck's disposal. (Nor could he find any scientific support for the company's suggestion that the results reflected Aleve's previously unrecognized protective powers rather than the dangers of Vioxx.)

"Deb had gone onto the FDA Web site to look at all the data presented in the advisory committee meetings—something, by the way, that I had never done," Topol said, shaking his head with a wan smile of admiration for the diligence of his young colleague. Vioxx was not the only cox-2 inhibitor on the market; Celebrex, made by Pfizer, was introduced that same year, and Bextra had also recently been approved by the FDA. But each drug functioned in a slightly different way and the numbers for Vioxx were more disturbing than those for either of the others. Mukherjee told Topol that there was a "real problem with Vioxx in particular," he recalled. "I had many other things on my mind so I said, 'Nahhh, it's not that big a deal. We have different work to do. Let's not waste our time on this.' Deb wouldn't have it. He insisted that I look at the data with him, absolutely insisted. So I did."

Once Topol had the statistics in front of him, he saw why

Mukherjee had become so agitated. "The evidence was right there," he said. "I still cannot believe that nobody else had seen it. That's when I began to understand what was really going on in that *USA Today* story. It just clicked: the company was attributing these miraculous functions to Aleve instead of investigating the potential dangers of their new drug. They were playing games: what they said at the time didn't seem to be an outright lie, but it also wasn't the truth that people needed to know. I said let's write this up. After all, these data mattered. It wasn't even a heart study, it was supposed to assess stomach complications, but you can't just shy away from information like that. There were too many lives at stake."

Topol and Mukherjee quickly put a paper together, along with Steven Nissen, another prominent cardiologist at the Cleveland Clinic, who had attended the advisory meeting where Vioxx was approved. "Deb drove the research and I gave it a framework," Topol said. The paper was the first independent analysis to include *all* the data the FDA had obtained from the VIGOR study, and it cast serious doubt on the supposition that naproxen offered special cardiovascular protection. The study was published in the *Journal of the American Medical Association* later that year. The three stopped short of calling for a moratorium on the use of Vioxx—the data at their disposal were not conclusive enough to warrant such a suggestion. However, they did warn doctors to take special care when prescribing the drug to people with heart disease. In their review, the authors stressed that Vioxx and other cox-2 inhibitors could cause serious side effects, and that a broader examination of their impact would be essential. "Given the remarkable exposure and popularity of this new class of medications," they wrote, "we believe that it is mandatory to conduct a trial specifically assessing cardiovascular risk and benefit of these agents. Until then, we urge caution in prescribing these agents to patients at risk for cardiovascular morbidity."

ERIC TOPOL IS TANNED and trim, a gangly man in his fifties with an oval face, graying hair that has begun to thin, and the type of relaxed affect that only someone who moves from a climate like Cleveland's to balmy San Diego could cultivate. Today, Topol has an entirely new kind of job: he is professor of genomics and the director of the Scripps Translational Science Institute in La Jolla. Scripps, one of the nation's largest biomedical research organizations, is eager to apply the emerging science of genomics—the information contained within our genes—to clinical medicine. Topol believes genetics will soon provide the knowledge we need to make substantial reductions in the incidence of heart disease. And that knowledge, of genetic predispositions and their implications for individuals, is increasing rapidly. Naturally, any reduction in the rate of heart disease, which kills at least a million Americans every year, would have a profound impact on public health. The field is young, and it hasn't been long since Scripps decided to invest heavily in it. When I visited La Jolla in the spring of 2008, the institute's building was only partially finished. Several floors consisted of little more than concrete shells and plastic sheeting. The setting, though, was spectacular: Topol's office looks out onto the Torrey Pines Golf Course, and beyond that, the Pacific Ocean. As I watched from his office window, dozens of people floated by on parasails before gently setting down in the shimmering green sea.

If Topol's life seems enviable, it hasn't been that way for long. His persistent criticism of Merck and, by implication, of the FDA, lasted three years, during which time Vioxx killed thousands of people. Topol found himself an outcast in his own profession, shunned for his warnings and eventually driven from the department he made famous. "There were years of sleepless nights, of bitterness," he said, speaking almost as if he were describing the ordeal of another person. "Years during which I allowed myself to wonder what on earth were we doing to people in the name of science." He came to realize that Merck possessed data that should have led the company to question Vioxx's safety long before it was approved.

Research published in the British medical journal the *Lancet* has estimated that 88,000 Americans had heart attacks after taking Vioxx, and that 38,000 of them died. In testimony before Congress, David Graham, the FDA's senior drug safety researcher, said the death toll may have reached as high as 55,000—almost exactly the number of American soldiers killed in the Vietnam War. (We will never know for sure; there is no precise way to account for the number of deaths caused by a drug like Vioxx, which was taken by millions of people. It's easy to notice an increased rate of heart attacks in a large group. Proving with certainty the specific cause of any one of them is nearly impossible.)

"This was one of the most remarkable breaches of trust in American scientific history," Topol said. He stood up and poured coffee. "It has been years now and even today nothing fundamental has occurred to ensure that this won't happen again. That is what amazes me. It is as if we took nothing from this tragedy but fear and a sense that those scientists were liars. One of the most important drugs of our time was released without the proper safety checks, and after the company's own scientists wondered in writing whether it would kill people. When Americans say they are skittish about the drug system and about conventional medicine itself, can anyone really be surprised?"

There have been a raft of studies released about Vioxx—from the Government Accountability Office, the Institute of Medicine, and many private organizations; eventually both the Senate Finance Committee and the House Committee on Government Reform held hearings. Every report, and much of the testimony, described the FDA's bureaucratic inefficiencies, its reluctance to take controversial positions, and Merck's willingness to exploit those weaknesses. "The FDA set up an internal safety auditing group for drugs going onto the market," Topol said. "They have responded by approving fewer new drugs than at any time in their history. The organization is just paralyzed."

By the end of the millennium, Americans had come to assume that a drug approved by the federal government was a drug they could swallow without wondering whether it would kill them. Vioxx changed all that. Thousands of lawsuits later, in 2007, the company placed nearly $5 billion into a settlement fund. That deposit permitted Merck to avoid nearly fifty thousand lawsuits. It also brought an end to hundreds of class-action cases filed on behalf of dead or injured Vioxx users, which, had they succeeded, could well have put Merck out of business. The settlement was the largest ever made by a pharmaceutical company. As part of the deal, Merck was never forced to admit fault in a single one of those deaths.

The Vioxx episode wove strands of fear and uncertainty together with an inchoate sense, shared by large swaths of American society, that we are ceding control of our lives to technology, particularly to highly sophisticated technology we can barely understand, and that we are doing so at a speed that seems to accelerate every year. Denialism is at least partly a defense against that sense of helplessness. What person, after watching Vioxx kill her husband, wouldn't say no to the next wonder drug? The story line—a predatory drug company lusting for profits—is not entirely new. The notion of technology as a force that does more harm than good, and of scientists toying with human life, dates back at least to Shelley and Goethe. Rousseau, the first Romantic, longed for the innocence and supposed simplicity of nature. He was convinced that science would have a pernicious effect on society, promising more than it could possibly deliver. G. K. Chesterton, in his book *Eugenics and Other Evils*, was even more direct, referring to organized science as "government tyranny."

It would be hard to find many examples in the past four centuries of scientists acting together to threaten humanity. Only a few are necessary, though, and there have been enough dark moments along the remarkable march of progress to generate anxiety and feed denial. Most of those moments were caused by error, not evil. Yes, in 1970 the Ford Motor Company, probably the definitive industrial symbol of twentieth-century America, introduced a car, the Pinto, that its engineers knew was likely to kill passengers. (Before the Pinto was introduced, in what may well stand out as the most remarkable memo in the history of engineering, Ford statisticians argued that the $11 cost of fixing each car added up to more than twice the amount of money—at $200,000 per

burn death and $67,000 for each serious injury—that they would
have had to pay in lawsuits or settlements.)

More often, there are simpler reasons to question the primacy
of science and technology. In the name of improving our lives,
some of the smartest people on earth have managed to ruin quite
a few of them. DES, or diethylstilbestrol, was the first synthetic
estrogen. It was cheap, easy to produce, and unusually potent.
First prescribed in 1938, DES was given to women who had experienced miscarriages or premature deliveries. Despite mixed
results in the laboratory, the drug was considered safe and effective
both for a pregnant woman and her developing fetus. It wasn't.
In the United States, as many as ten million people were exposed
to DES by 1971, when it was pulled from the market. "DES
Daughters," as they came to be known, are at increased risk for
several types of cancer, as well as structural abnormalities of the
reproductive tract, complications with pregnancy, and infertility.

Fear builds far more easily than it dissipates. Ronald Reagan
once famously claimed that the "nine most terrifying words in the
English language are: 'I'm from the government and I'm here
to help.'" If anyone needed a reminder of how far our faith in
science had fallen by the end of the twentieth century, Vioxx demonstrated that five other words could prove just as frightening:
"Trust me, I'm a scientist." It was quite a crash. Pharmaceutical
companies were among the most highly valued institutions in
America after World War II, and it's not hard to see why. They
introduced the core values of consumer culture to American
medicine. Drugs became like everything else Made in America:
products meant to ease life and solve problems.

The flood of new antibiotics and the rapid development of vaccinations for everything from diphtheria to polio helped define the spirit of the country with a single word: optimism. America was a can-do nation and it possessed technology that would solve the world's problems. From infectious disease to cancer, and from pollution to hunger, we would overcome it all. Nylon, Lycra, Teflon, Kevlar, and Mylar, for example—all made by DuPont—were all triumphs of ease and modernity. We could fix whatever was broken, cure whatever ailed us, and make life easier for everyone along the way. Today the words "DuPont," "Merck," and "Monsanto" are often used as epithets, and they compete with tobacco companies for the role of the most loathed American corporations. Conventional medicine and technology itself, despite their clear rates of success, seem to many people as likely to cause danger as to enhance our lives.

In a Harris poll of attitudes toward corporate America published in 2008, just 27 percent of respondents said they "somewhat or strongly" trusted the pharmaceutical industry. More than half described their views as firmly negative, which places Big Pharma slightly below Big Oil, and a bit above tobacco companies, in the esteem of the average American. The figures for the FDA were only marginally higher. What was written with sincerity only a few decades ago is now played strictly for laughs: in 2006, the satirical newspaper the *Onion* ran a story in its "Science and Technology" section under this headline: "Wonder Drug Inspires Deep, Unwavering Love of Pharmaceutical Companies." The year before, a film version of John le Carré's novel *The Constant Gardener* was released. Like the book, it portrayed an international pharmaceu-

tical conglomerate as avaricious and cartoonishly evil. The plot
was ludicrous and appealed to the worst possible stereotypes of
mindless capitalism. But people ate it up.

Vioxx and other preventable catastrophes ensured that they
would. Corporations, wrapping themselves in the mantle of progress but all too often propelled by greed, have done more than
religion or even Luddism to inflame denialists and raise doubts
about the objectivity of science. In 2008, reports surfaced that for
more than a year, Merck and Schering-Plough concealed the
fact that their jointly marketed cholesterol drug, Vytorin, was no
more effective than generic statins costing less than half as much.
Nonetheless, the companies still spent more than $100 million
during that time to advertise the drug's special qualities.

News like that has become routine. In early 2009, the British
company AstraZeneca was found to have somehow mislaid unfa-
vorable studies about its antipsychotic drug Seroquel. At about the
same time, more than one hundred students at the Harvard Medical School publicly questioned the ethics of their professors, some
of whom are frequently paid as consultants by the pharmaceutical
companies whose products they are supposed to judge. The situation got so bad that the Institute of Medicine, the branch of the
National Academy of Sciences charged with providing advice on
critical issues of medical research, denounced physicians who
accepted money from drug companies. "It is time for medical
schools to end a number of long-accepted relationships and practices that create conflicts of interest, threaten the integrity of their
missions and their reputations, and put public trust in jeopardy,"
the IOM concluded in a report. Just two weeks later, the *Scientist*
revealed that Merck had published a journal filled with favorable

articles about more than one of the company's drugs, without ever bothering to disclose that the publication, the *Australasian Journal of Bone and Joint Medicine*, was sponsored by the company itself. "To the jaundiced eye, [the journal] might be detected for what it is: marketing," Public Citizen's Peter Lurie said. "Many doctors would fail to identify that and might be influenced by what they read."

If all that wasn't sufficiently damning, the Baystate Medical Center in Springfield, Massachusetts, revealed that Scott S. Reuben, its highly influential former physician in charge of acute pain treatment, fabricated data from twenty-one medical studies that claimed to show the benefits of painkillers like Vioxx and Celebrex. "The pharmas are in big trouble in terms of credibility," said Rob Frankel, a brand consultant who focuses on medical industries. "They're just above Congress and used-car salesmen."

THIRTY YEARS AGO NOBODY discussed the principal motive behind scientific research: nobody needed to. It was a quest for knowledge. Today, the default assumption is that money matters most of all, and people tend to see science through the prism of commerce. At least until Viagra was introduced, and endorsed on television by Bob Dole, a former candidate for the presidency, no drug had been marketed more successfully than Vioxx. In 2000, the year after it first appeared, Merck spent $160 million advertising their painkiller. They were able to do that thanks to the advent, just three years before, of direct-to-consumer advertising. Only two countries allowed pharmaceutical companies to advertise prescription drugs directly to consumers: New Zealand and the United States.

In America, such ads virtually always consist of glossy promotional materials used to announce major medical advances. (The federal government requires that they include tiny print "information" presented in medical jargon, the meaning of which, for most consumers, is nearly impossible to understand.) These advertisements are not really intended to educate patients, nor to help them become more sophisticated about their own health. They are purely an attempt to get doctors to fill more prescriptions, and they work with stunning regularity. "Blockbuster" drugs like Vioxx, Viagra, and the cholesterol medicine Lipitor can become a multinational corporation's central source of income.

Our regulatory system encourages companies to invest in marketing, not in research: in the United States, a new drug typically takes a decade to develop and costs hundreds of millions of dollars. With stakes that high, and lawsuits waiting for any company that commits even the smallest error, pharmaceutical firms are far more likely to profit from aggressive sales of products that are already available than from introducing anything new. One reason the ads succeed is that it is nearly impossible to spend a day without seeing one or more major drugs advertised on television. (Another reason is the amount of money companies spend. The marketing budget for AstraZeneca's heartburn pill Nexium, a sure moneymaker, is bigger than the comparable budget for Budweiser beer.)

Rheumatoid arthritis afflicts more than two million Americans, and it can be devastating. Many of those people were overjoyed when advertisements for Vioxx began blanketing the airwaves in 1999. Suddenly, men who couldn't bend over were tying their shoes again, walking dogs, and regaining lives that slowly had been consumed by pain. Dorothy Hamill skated across millions of television screens on behalf of Vioxx, overcoming arthritis and moving with the nimble certainty of a teenager at the Olympics. "People dancing in the streets, twirling their partners in joy," Topol said. "That's all anybody ever saw." Go talk to your doctor about Vioxx, the ad would say. And people did, by the millions. In 1996, American pharmaceutical companies spent $11.4 billion on direct advertisements; by 2005 the figure was more than $29 billion. Doctors were overwhelmed with requests, and for the most part were only too happy to comply, writing nearly one hundred million prescriptions for Vioxx alone between 1999 and 2004.

The words "ask your doctor" have become code for "change your prescription." Often, what people ended up with was no better than the cheaper and more readily available drugs they were taking in the first place. And just as often, the drugs for sale were so new to the market that their safety was hard to gauge. "Americans must face an inconvenient truth about drug safety," Henry Waxman, the veteran California congressman, said when asked his position on the impact of these advertisements. Waxman is perhaps the most astute congressional observer of American medicine, and since the election of Barack Obama, in his new role as chairman of the House Energy and Commerce Committee, he may also be the most powerful. "The truth is that we inevitably allow drugs on the market whose risks are not fully known," he said. In 2006, the Institute of Medicine suggested a moratorium on such advertising, a brief pause before permitting companies like Merck to hawk powerful chemicals as if they were Cheerios or vacuum cleaners. That would certainly have saved many of the lives lost to Vioxx.

People tend to see what they are looking for, however, and to millions, pain relief was all that mattered. When it comes to getting a message across, the Super Bowl, where ads for drugs like Vioxx and Viagra have been ubiquitous, will always matter more than the *New England Journal of Medicine*. In this atmosphere, Eric Topol was largely on his own, and he became the target of Merck's fullest fury. "Merck went after me with all they had," he told me. Topol had collaborated with Merck often—at the time of his clash with the company over Vioxx he was actually running one of its trials, for the anti-platelet drug Aggrastat. Like most people in his profession, Topol considered Merck a remarkable place. During one stretch beginning in 1987, it was named by *Fortune* magazine for seven straight years as the most admired corporation in America, a record that remains unmatched. Merck seemed to prove that profits and decency were not incompatible. "I was in no way predisposed to have negative feelings toward Merck," Topol said. "In fact it was quite the contrary." None of that mattered, though, because they came after him with a cleaver. "They said my first paper was 'data dredging,'" by which they meant a pedantic report full of numbers that proved nothing. "They told anyone who would listen that I had been a fine researcher until the day that paper was published, but at that moment I had suddenly lost it. And at the time all I had said was, 'Wait a minute, there needs to be some more study here.'"

By the end of 2001, however, Topol had moved on. He had begun to focus primarily on genomics and his research switched from treating heart attacks to preventing them. Others were studying whether Vioxx increased the risk of heart attacks, and he felt he had done all he could to address the issue. At the time, Topol had no idea how divisive Vioxx had become at Merck itself. It turned out that scientists there had worried as early as 1996 about the effect the drug would have on the cardiovascular system. The reason for that concern was clear. Vioxx altered the ratio of two crucial substances, the hormone prostacyclin and a molecule called thromboxane, which together help balance blood flow and its ability to clot properly. Suppressing prostacyclin reduces inflammation and pain, and that made Vioxx work. Suppress it too powerfully, however, and thromboxane can cause increased blood pressure and too much clotting, either of which can lead to heart attacks. By 2002, Merck decided to embark on a major study of the cardiovascular risks caused by Vioxx—just as Topol and his colleagues had suggested. The trial would have produced useful data fairly rapidly, but just before it began, the company abruptly scuttled the project. In the end, Merck never made any significant effort to assess the cardiovascular risk posed by its most successful product.

Instead, the company issued a "Cardiovascular Card" to sales representatives. More than three thousand members of the sales force were instructed to refer doctors with questions to the card, which claimed—falsely—that Vioxx was eight to eleven times safer than other similar painkillers. The sales reps were told to produce the card only when all else had failed, to help "physicians in response to their questions regarding the cardiovascular effects of Vioxx." The FDA, realizing that doctors needed to understand the gravity of the findings from the VIGOR trial, issued a strongly worded letter instructing Merck to correct the record. "Your claim in the press release that Vioxx has a 'favorable cardiovascular safety profile,'" the letter read in part, "is simply incomprehensible, given the rate of heart attack and serious cardiovascular events compared to naproxen." The company reacted swiftly: "Do not initiate discussions of the FDA arthritis committee . . . or the results of the . . . VIGOR study," the sales force was told. If doctors proved too querulous, the Merck representatives were instructed to respond by saying, "I cannot discuss the study with you."

In the summer of 2004, the *Lancet* asked Topol and the gastroenterologist Gary W. Falk, a colleague of Topol's from the Cleveland Clinic, to sum up the state of knowledge about Vioxx and similar drugs known, as a class, by the name coxibs. Their editorial was published that August under the title "A Coxib a Day Won't Keep the Doctor Away." Topol was taken aback when he realized how little had changed. "It was amazing to see that nothing had been done in three years," he recalled. "It was not even clear that Vioxx protected the stomach. It cost four dollars a day for these darn pills." On September 29 of that year, Topol happened to dine with Roy Vagelos, Merck's much-admired former chief executive, who had been retired for nearly a decade. Topol was visiting a New York–based biopharmaceutical company called Regeneron whose board Vagelos chairs. There were few people in medicine for whom Topol had as much respect. "We started talking about Vioxx," he said. "It was the first time I ever spoke to Roy about it. I remember that conversation well: it was at Ruth's Chris Steak House in Westchester. Roy went on for a while. He was entirely opposed to the Merck approach to Vioxx. And he didn't mince words."

The following morning a Merck cardiologist told him the company was removing Vioxx from other trial had shown that patients taking the drug risk of heart attack and stroke. That study, APPROVe, began 2000 as an attempt to discover whether Vioxx helped prevent the recurrence of colon polyps. It didn't. "I was shocked," Topol said. "But I thought that it was responsible for them to pull it. And Steve Nissen came down to my office, also very pleased, and said, 'Isn't that great? They are pulling the drug.' We both thought it was the right thing to do."

For Topol, that could well have been how the story ended. But Merck began to mount a press offensive. The message never varied: Merck put patients first. "Everything they had ever done in the course of Vioxx was putting patients first. All the data was out there," Topol said, still stunned by the brazen public lies. "This just wasn't true. It wasn't right. I called and tried to speak to Ray Gilmartin"— Merck's chief executive. "Neither he nor anyone else returned my calls." (That itself was significant: after all, Topol ran one of the most important cardiology departments in the country; he was also the director of a Merck drug trial.) "This was a breach of trust that really rocked the faith people have in institutions like those," Topol said. "We are talking about thousands of heart attacks. There were simply gross discrepancies in what they presented to the FDA and what was published in journals. I took them on. I had to."

That week, Topol wrote an op-ed piece for the *New York Times*. He called it "Vioxx Vanished." The *Times* had a better idea: "Good Riddance to a Bad Drug." Noting that Vioxx increased the risk of heart attacks and strokes, Topol wrote that "our two most mmon deadly diseases should not be caused by a drug." He also published a column in the *New England Journal of Medicine*, called "Failing the Public Health": "The senior executives at Merck and the leadership at the FDA," he wrote, "share responsibility for not having taken appropriate action and not recognizing that they are accountable for the public health."

On December 3, 2005, in a videotaped deposition presented under subpeona at one of the many trials following the recall, Topol argued that Vioxx posed an "extraordinary risk." A colleague from the Cleveland Clinic, Richard Rudick, told him that Gilmartin, the Merck CEO, had become infuriated by Topol's public attacks and had complained bitterly to the clinic's board about the articles in the *Times* and the *New England Journal of Medicine*. "What has Merck ever done to the clinic to warrant this?" Gilmartin asked.

Two days after that testimony, Topol received an early call telling him not to attend an 8 a.m. meeting of the board of governors. "My position—chief academic officer—had been abolished. I was also removed as provost of the medical school I founded." The clinic released a statement saying that there was no connection between Topol's Vioxx testimony and his sudden demotion, after fifteen years, from one of medicine's most prominent positions. A spokeswoman for the clinic called it a simple reorganization. The timing, she assured reporters, was a coincidence.

DID THE RECALL of Vioxx, or any other single event, cause millions of Americans to question the value of science as reflexively as they had once embraced it? Of course not. Over the decades, as our knowledge of the physical world has grown, we have also endured the steady drip of doubt—about both the definition of progress and whether the pursuit of science will always drive us in the direction we want to go. A market disaster like Vioxx, whether through malice, greed, or simply error, presented denialists with a rare opportunity: their claims of conspiracy actually came true. More than that, in pursuit of profits, it seemed as if a much-admired corporation had completely ignored the interests of its customers.

It is also true, however, that spectacular technology can backfire spectacularly—and science doesn't always live up to its expectations. When we see something fail that we had assumed would work, whether it's a "miracle" drug or a powerful machine, we respond with fear and anger. People often point to the atomic bomb as the most telling evidence of that phenomenon. That's not entirely fair: however much we may regret it, the bomb did what it was invented to do.

That wasn't the case in 1984, when a Union Carbide pesticide factory in Bhopal, India, released forty-two tons of toxic methyl isocyanate gas into the atmosphere, exposing more than half a million people to deadly fumes. The immediate death toll was 2,259; within two weeks that number grew to more than eight thousand. Nor was it true two years later, when an explosion at Unit 4 of the V. I. Lenin Atomic Power Station transformed a place called Chernobyl into a synonym for technological disaster. They were the worst industrial accidents in history—one inflicting immense casualties and the other a worldwide sense of dread. The message was hard to misinterpret: "Our lives depend on decisions made by other people; we have no control over these decisions and

usually we do not even know the people who make them," wrote
Ted Kaczynski, better known as the Unabomber, in his essay "Industrial Society and Its Future"—the Unabomber Manifesto. "Our
lives depend on whether safety standards at a nuclear power plant
are properly maintained; on how much pesticide is allowed to
get into our food or how much pollution into our air; on how
skillful (or incompetent) our doctor is. . . . The individual's
search for security is therefore frustrated, which leads to a sense of
powerlessness."

Kaczynski's actions were violent, inexcusable, and antithetical
to the spirit of humanity he professed to revere. But who hasn't felt
that sense of powerlessness or frustration? Reaping the benefits
of technology often means giving up control. That only matters,
of course, when something goes wrong. Few of us know how to fix
our carburetors, or understand the mechanism that permits telephone calls to bounce instantly off satellites orbiting twenty-eight
thousand miles above the earth only to land a split second later in
somebody else's phone on the other side of the world.

That's okay; we don't need to know how they function, as long
as they do. Two hundred or even fifty years ago, most people
understood their material possessions—in many cases they created
them. That is no longer the case. Who can explain how their
computer receives its constant stream of data from the Internet?
Or understands the fundamental physics of a microwave? When
you swallow antibiotics, or give them to your children, do you
have any idea how they work? Or how preservatives are mixed
into many of the foods we eat or why? The proportion of our
surroundings that any ordinary person can explain today is
minute—and it keeps getting smaller.

This growing gap between what we do every day and what we know how to do only makes us more desperate to find an easy explanation when something goes wrong. Denialism provides a way to cope with medical mistakes like Vioxx and to explain the technological errors of Chernobyl or Bhopal. There are no reassuring safety statistics during disasters and nobody wants to hear about the tens of thousands of factories that function flawlessly, because triumphs are expected, whereas calamities are unforgettable. That's why anyone alive on January 28, 1986, is likely to remember that clear, cold day in central Florida, when the space shuttle *Challenger* lifted off from the Kennedy Space Center, only to explode seventy-three seconds later, then disintegrate in a dense white plume over the Atlantic. It would be hard to overstate the impact of that accident. The space program was the signature accomplishment of American technology: it took us to the moon, helped hold back the Russians, and made millions believe there was nothing we couldn't accomplish. Even our most compelling disaster—the Apollo 13 mission—was a successful failure, ending with the triumph of technological mastery needed to bring the astronauts safely back to earth.

By 1986, America had become so confident in its ability to control the rockets we routinely sent into space that on that particular January morning, along with its regular crew, NASA strapped a thirty-seven-year-old high school teacher named Christa McAuliffe from Concord, New Hampshire, onto what essentially was a giant bomb. She was the first participant in the new Teacher in Space program. And the last.

The catastrophe was examined in merciless detail at many nationally televised hearings. During the most remarkable of

them, Richard Feynman stunned the nation with a simple display
of show-and-tell. Feynman, a no-nonsense man and one of the
twentieth century's greatest physicists, dropped a rubber O-ring
into a glass of ice water, where it quickly lost resilience and cracked.
The ring, used as a flexible buffer, couldn't take the stress of the
cold, and it turned out neither could one just like it on the shuttle
booster rocket that unusually icy day in January. Like so many of
our technological catastrophes, this was not wholly unforeseen.
"My God, Thiokol, when do you want me to launch, next April?"
Lawrence Mulloy, manager of the Solid Rocket Booster Project at
NASA's Marshall Space Flight Center, complained to the manufacturer, Morton Thiokol, when engineers from the company
warned him the temperature was too low to guarantee their product would function properly.

SCIENTISTS HAVE NEVER BEEN good about explaining what they
do or how they do it. Like all human beings, though, they make
mistakes, and sometimes abuse their power. The most cited of
those abuses are the twins studies and other atrocities carried out
by Nazi doctors under the supervision of Josef Mengele. While
not as purely evil (because almost nothing could be), the most
notorious event in American medical history occurred not long
ago: from 1932 to 1972, in what became known as the Tuskegee
Experiment, U.S. Public Health Service researchers refused to
treat hundreds of poor, mostly illiterate African American share-
croppers for syphilis in order to get a better understanding of the
natural progression of their disease. Acts of purposeful malevo-

lence like those have been rare; the more subtle scientific tyranny of the elite has not been.

In 1883, Charles Darwin's cousin Francis Galton coined the term "eugenics," which would turn out to be one of the most heavily freighted words in the history of science. Taken from a Greek word meaning "good in birth," eugenics, as Galton defined it, simply meant improving the stock of humanity through breeding. Galton was convinced that positive characteristics like intelligence and beauty, as well as less desirable attributes like criminality and feeblemindedness, were wholly inherited and that a society could breed for them (or get rid of them) as they would, say, a Lipizzaner stallion or a tangerine. The idea was that with proper selection of mates we could dispense with many of the ills that kill us—high blood pressure, for instance, or obesity, as well as many types of cancer.

Galton saw this as natural selection with a twist, and felt it would provide "the more suitable races or strains of blood a better chance of prevailing speedily over the less suitable." Galton posed the fundemental question in his 1869 book, *Hereditary Genius*: would it not be "quite practicable to produce a highly gifted race of men by judicious marriages during consecutive generations?" As the Yale historian Daniel J. Kevles points out in his definitive 1985 study *In the Name of Eugenics*, geneticists loved the idea and eagerly attempted to put it into action. Eugenics found particular favor in the United States.

In 1936, the American Neurological Association published a thick volume titled *Eugenical Sterilization*, a report issued by many of the leading doctors in the United States and funded by the Carnegie Foundation. There were chapters on who should be sterilized and who shouldn't, and it was chock full of charts and scientific data—about who entered the New York City hospital system, for example, at what age and for what purpose. The board of the ANA noted in a preface that "the report was presented to and approved by the American Neurological Association at its annual meeting in 1935. It had such an enthusiastic reception that it was felt advisable to publish it in a more permanent form and make it available to the general public."

Had their first recommendation appeared in a novel, no reader would have taken it seriously: "Our knowledge of human genetics has not the precision nor amplitude which would warrant the sterilization of people *who themselves are normal* [italics in the original] in order to prevent the appearance, in their descendants, of manic-depressive psychosis, dementia praecox, feeblemindedness, epilepsy, criminal conduct or any of the conditions which we have had under consideration. An exception may exist in the case of normal parents of one or more children suffering from certain familial diseases, such as Tay-Sachs' amaurotic idiocy." Of course, for people who were not considered normal, eugenics had already arrived. Between 1907 and 1928, nearly ten thousand Americans were sterilized on the general grounds that they were feebleminded. Some lawmakers even tried to make welfare and unemployment relief contingent upon sterilization.

Today, our knowledge of genetics has both the precision and the amplitude it lacked seventy years ago. The Nazis helped us bury thoughts of eugenics, at least for a while. The subject remains hard to contemplate—but eventually, in the world of genomics, impossible to ignore. Nobody likes to dwell on evil. Yet there has never been a worse time for myopia or forgetfulness. By forgetting the Vioxxes, Vytorins, the nuclear accidents, and constant flirtation with eugenics, and instead speaking only of science as a vehicle for miracles, we dismiss an important aspect of who we are. We need to remember both sides of any equation or we risk acting as if no mistakes are possible, no grievances just. This is an aspect of denialism shared broadly throughout society; we tend to consider only what matters to us now, and we create expectations for all kinds of technology that are simply impossible to meet. That always makes it easier for people, already skittish about their place in a complex world, to question whether vaccines work, or AIDS is caused by HIV, or why they ought to take prescribed pain medication instead of chondroitin or some other useless remedy recommended wholeheartedly by alternative healers throughout the nation.

IF YOU LIVED with intractable pain, would you risk a heart attack to stop it? What chances would be acceptable? One in ten? One in ten thousand? "These questions are impossible to answer completely," Eric Topol told me when I asked him about it one day as we walked along the beach in California. "Merck sold Vioxx in an unacceptable and unethical way. But I would be perfectly happy if it was back on the market."

Huh? Eric Topol endorsing Vioxx seemed to make as much sense as Alice Waters campaigning for Monsanto and genetically modified food. "I can't stress strongly enough how deplorable this catastrophe has been," he said. "But you have to judge risk properly and almost nobody does. For one thing, you rarely see a

discussion of the effect of *not* having drugs available." Risk always
has a numerator and a denominator. People tend to look at only
one of those numbers, though, and they are far more likely to
remember the bad than the good. That's why we can fear flying
although it is hundreds of times safer than almost any other form
of transportation. When a plane crashes we *see* it. Nobody comes
on television to announce the tens of thousands of safe landings
that occur throughout the world each day.

We make similar mistakes when judging our risks of illness.
Disease risks are almost invariably presented as statistics, and what
does it mean to have a lifetime heart attack risk 1.75 times greater
than average? Or four times the risk of developing a certain kind
of cancer? That depends: four times the risk of developing a cancer
that affects 1 percent of the population isn't terrible news. On the
other hand, a heart attack risk 75 percent greater than average, in
a nation where heart attacks are epidemic, presents a real problem.
Few people, however, see graphic reality in numbers. We are simply
not good at processing probabilistic information. Even something
as straightforward as the relationship between cigarette smoking and cancer isn't all that straightforward. When you tell a
smoker he has a 25 percent chance of dying from cancer, the natural response is to wonder, "From *this* cigarette? And how likely is
that really?" It is genuinely hard to know, so all too often we let
emotion take over, both as individuals and as a culture.

The week in 2003 that SARS swept through Hong Kong, the
territory's vast new airport was deserted, and so were the city's
usually impassable streets. Terrified merchants sold face masks and
hand sanitizer to anyone foolish enough to go out in public. SARS
was a serious disease, the first easily transmitted virus to emerge

in the new millennium. Still, it killed fewer than a thousand people, according to World Health Organization statistics. Nevertheless, "it has been calculated that the SARS panic cost more than $37 billion globally," Lars Svendsen wrote in *A Philosophy of Fear*. "For such a sum one probably could have eradicated tuberculosis, which costs several million people's lives every year."

Harm isn't simply a philosophical concept; it can be quantified. When Merck, or any another company, withholds information that would have explained why a drug might "fail," people have a right to their anger. Nonetheless, the bigger problem has little to do with any particular product or industry, but with the way we look at risk. America takes the Hollywood approach, going to extremes to avoid the rare but dramatic risk—the chance that minute residues of pesticide applied to our food will kill us, or that we will die in a plane crash. (There is no bigger scam than those insurance machines near airport gates, which urge passengers to buy a policy just in case the worst happens. A traveler is more likely to win the lottery than die in an airplane. According to Federal Aviation Administration statistics, scheduled American flights spent nearly nineteen million hours in the air in 2008. There wasn't one fatality.)

On the other hand, we constantly expose ourselves to the likely risks of daily life, riding bicycles (and even motorcycles) without helmets, for example. We think nothing of exceeding the speed limit, and rarely worry about the safety features of the cars we drive. The dramatic rarities, like plane crashes, don't kill us. The banalities of everyday life do.

We certainly know how to count the number of people who died while taking a particular medication, but we also ought to measure the deaths and injuries caused when certain drugs are *not* brought to market; that figure would almost always dwarf the harm caused by the drugs we actually use. That's even true with Vioxx. Aspirin, ibuprofen, and similar medications, when used regularly for chronic pain, cause gastrointestinal bleeding that contributes to the death of more than fifteen thousand people in the United States each year. Another hundred thousand are hospitalized. The injuries—including heart attacks and strokes—caused by Vioxx do not compare in volume. In one study of twenty-six hundred patients, Vioxx, when taken regularly for longer than eighteen months, caused fifteen heart attacks or strokes per every one thousand patients. The comparable figure for those who received a placebo was seven and a half per thousand. There was no increased cardiovascular risk reported for people who took Vioxx for less than eighteen months. In other words, Vioxx increased the risk of having a stroke or heart attack by less than 1 percent. Those are odds that many people might well have been happy to take.

"All Merck had to do was acknowledge the risk, and they fought that to the end," Topol said. "After fifteen months of haggling with the FDA they put a tiny label on the package that you would need a microscope to find. If they had done it properly and prominently, Vioxx would still be on the market. But doctors and patients would know that if they had heart issues they shouldn't take it."

Most human beings don't walk out the door trying to hurt other people. So if you are not deliberately trying to do harm, what are the rules for using medicine supposed to be? What level of risk would be unacceptable? A better question might be, Is any risk acceptable? Unfortunately, we have permitted the development of unrealistic standards that are almost impossible to attain. The pharmaceutical industry, in part through its own greed (but only in part), has placed itself in a position where the public expects it never to cause harm. Yet, drugs are chemicals we put into our body, and that entails risks. No matter how well they work, however, if one person in five thousand is injured, he could sue and have no trouble finding dozens of lawyers eager to represent him. People never measure the risk of keeping the drug off the market, though, and that is the problem. If you applied FDA phase I or II or III criteria—all required for drug approval—to driving an automobile in nearly any American city, nobody would be allowed to enter one. When we compare the risk of taking Vioxx to the risk of getting behind the wheel of a car, it's not at all clear which is more dangerous.

Vaccines and the
Great Denial

Marie McCormick is a studious and reserved woman with the type of entirely unthreatening demeanor that comes in handy in her job as a professor of pediatrics at the Harvard School of Public Health. She has devoted most of the past four decades to preparing physicians to nurture mothers and their children, and, since her days as a student at Johns Hopkins, has focused much of her research on high-risk newborns and infant mortality. Like many prominent academic physicians, her renown had largely been restricted to her field. Until 2001.

That was the year she was asked to lead a National Academy of Sciences commission on vaccine safety. The Immunization Safety Review Committee was established by the Institute of Medicine to issue impartial, authoritative, and scientifically rigorous reports on the safety of vaccinations. Its goal, while vital, seemed simple enough: bring clarity to an issue where too often confusion reigned. McCormick took on the assignment readily, although she was surprised at having been selected. It was not as if she considered vaccine safety unimportant—the issue had preoccupied her for decades. Nonetheless, vaccines were not McCormick's area of expertise and she couldn't help thinking that there must be someone better suited to the job. "My research has always been on the very premature," she explained. "So I was a bit naive about why they might want me to run that committee." She soon made a discovery that surprised her: "I realized that all of us on the committee were selected *because* we had no prior contact with vaccines, vaccine research, or vaccine policy. We all had very strong public health backgrounds, but we were just not clear about the nature or intensity of the controversy."

The controversy that the panel set out to address was whether the benefits of receiving childhood vaccines outweighed the risks. In particular, the committee was asked to investigate the suggested link between the measles, mumps, and rubella inoculation routinely administered between the ages of one and two and the development of autism, which often becomes apparent at about the same time. The incidence of autism has risen dramatically during the past three decades, from less than one child in twenty-five hundred in 1970 to nearly one in every 150 today. That amounts to fifty new diagnoses of autism or a related disorder every day—almost always in children who seem to be developing normally, until suddenly their fundamental cognitive and communication skills begin to slip away.

Parents, understandably desperate to find a cause and often

wholly unfamiliar with many diseases that vaccines prevent, began
to wonder—publicly and vocally—why their children even needed
them. There could be no better proof of just how effective those
vaccines have been. With the sole exceptions of improved sanita-
tion and clean drinking water, no public health measure has
enhanced the lives of a greater number of people than the
widespread adoption of vaccinations, not even the use of antibiotics. Cholera and yellow fever, both ruthless killers, are hardly
known now in the developed world. Until vaccines were discovered
to stop them, diphtheria and polio rolled viciously through America every year, killing thousands of children, paralyzing many
more, and leaving behind ruined families and a legacy of terror.
Both are gone. So is mumps, which in the 1960s infected a million
children every year (typically causing them to look briefly like
chipmunks, but occasionally infiltrating the linings of the brain
and spinal cord, causing seizures, meningitis, and death).

Even measles, an illness that most young parents have never
encountered, infected nearly four million Americans annually
until 1963, when a vaccine was introduced. Typically, hundreds
died, and thousands would become disabled for life by a condition
called measles encephalitis. (In parts of the developing world,
where vaccines are often unavailable, measles remains an unbridled
killer: in 2007, about two hundred thousand children died from
the disease—more than twenty every hour.) In the United States,
fifty-two million measles infections were prevented in the two
decades after the vaccine was released. Without the vaccine, seventeen thousand people would have been left mentally retarded,
and five thousand would have died. The economic impact has also
been dramatic: each dollar spent on the MMR vaccine saves nearly

twenty in direct medical costs. That's just money; in human terms, the value of avoiding the disease altogether cannot be calculated. By 1979, vaccination had even banished smallpox, the world's most lethal virus, which over centuries had wiped out billions of people, reshaping the demographic composition of the globe more profoundly than any war or revolution.

Those vaccines, and others, have prevented unimaginable misery. But the misery is only unimaginable to Americans today because they no longer need to know such diseases exist. That permits people to focus on risks they do confront, like those associated with vaccination itself. Those risks are minute, and side effects are almost always minor—swelling, for instance; a fever or rash. Still, no medical treatment is certain to work every time. And serious adverse reactions do occur. If you hunt around the Internet for an hour (or ten) you might think that nobody pays attention to vaccine safety in America today. The Public Health Service has actually never been more vigilant. For example, in 1999 the Centers for Disease Control called for an end to the use of the oral polio vaccine, developed by Albert Sabin, which, because it contained weakened but live virus, triggered the disease in about ten people out of the millions who took it each year. (A newer injectable and inactivated version eliminates even this tiny threat.) Despite legitimate concerns about safety, every vaccine sold in the United States is scrutinized by at least one panel of outside advisers to the Food and Drug Administration before it can be licensed; many don't even make it that far. As a result, vaccination for virtually every highly contagious disease is never as dangerous as contracting the infections those vaccines prevent.

Prevention is invisible, though, and people fear what they cannot see. Nobody celebrates when they avoid an illness they never expected to get. Humans don't think that way. Choosing to vaccinate an infant requires faith—in pharmaceutical companies, in public health officials, in doctors, and, above all, in science. These days, that kind of faith is hard to come by. So despite their success, there has been no more volatile subject in American medicine for the past decade than the safety of vaccines. There is a phrase used commonly in medicine: "true, true, and unrelated." It is meant to remind physicians not to confuse coincidence with cause. That kind of skepticism, while a fundamental tenet of scientific research, is less easily understood by laymen.

For most people, an anecdote drawn from their own lives will always carry more meaning than any statistic they might find buried in a government report. "Neither my husband nor anyone in his family has ever been vaccinated . . . and there isn't a single person in his family who has ever had anything worse than a cold," one woman wrote on the heavily read blog Mom Logic. "Myself and my family, on the other hand, were all vaccinated against every possible thing you could imagine. . . . Somehow we all got the flu every single year. Somehow everyone in my family is chronically ill. And amazingly, when the people in my family reach 50 they are all old and deteriorated. In my husband's family they are all vibrant into their late 90's. My children will not be vaccinated."

This particular epidemic of doubt began in Britain, when the *Lancet* published a 1998 study led by Dr. Andrew Wakefield in which he connected the symptoms of autism directly to the MMR vaccine. The study was severely flawed, has been thoroughly discredited, and eventually ten of its thirteen authors retracted their contributions. Yet the panic that swept through Britain was breathtaking: vaccination rates fell from 92 percent to 73 percent and in parts of London to nearly 50 percent. Prime Minister Tony Blair refused repeatedly to respond to questions about whether his youngest child, Leo, born the year after Wakefield's study, received the standard MMR vaccination. Blair said at the time that medical treatment was a personal matter and that inquiries about his children were unfair and intrusive. No virus respects privacy, however, so public health is *never* solely personal, as the impact on Britain has shown. England and Wales had more cases of measles in 2006 and 2007 than in the previous ten years combined. In 2008, the caseload grew again—this time by nearly 50 percent. The numbers in the United States have risen steadily as well, and the World Health Organization has concluded that Europe, which had been on track to eliminate measles by 2010, is no longer likely to succeed. Vaccination rates just aren't high enough.

Fear is more infectious than any virus, and it has permitted politics, not science, to turn one of the signature achievements of modern medicine into fodder for talk show debates and marches on Washington. Celebrities like Jenny McCarthy, who oppose the need for a standard vaccination schedule, denounce celebrities like Amanda Peet who are willing to say publicly that the benefits of vaccines greatly outweigh the risks. Peet represents Every Child by Two, a nonprofit organization that supports universal vaccination. Not long after she began speaking for the group, Peet and McCarthy began to clash. At one point, McCarthy reminded Peet that she was right because "there is an angry mob on my side." When three physicians, appearing on *Larry King Live*, disagreed with McCarthy, she simply shouted "Bullshit!" in response. When that didn't shut them up, she shouted louder. Data, no matter how solid or frequently replicated, seems beside the point.

What does it say about the relative roles that denialism and reason play in a society when a man like Blair, one of the democratic world's best-known and most enlightened leaders, refused at first to speak in favor of the MMR vaccine, or when a complete lack of expertise can be considered a *requirement* for participation in America's most prominent vaccine advisory commission? "Politically, there is simply no other way to do it," Anthony S. Fauci explained. "Experts are often considered tainted. It is an extremely frustrating fact of modern scientific life." Fauci has for many years run the National Institute of Allergy and Infectious Diseases, where at the beginning of the AIDS epidemic he emerged as one of the public health establishment's most eloquent and reliably honest voices. He shook his head in resignation when asked about the need for such a qualification, but noted that it has become difficult to place specialists on committees where politics and science might clash. "You bring people with histories to the table and they are going to get pummeled," he said. "It would simply be war."

War is exactly what the vaccine commission got. During McCormick's tenure, the National Academy of Sciences published several reports of its findings. In a 2001 study, *Measles-Mumps-Rubella Vaccine and Autism*, the committee concluded that there was no known data connecting MMR immunizations with the spectrum of conditions that are usually defined as autism. The wording left room for doubt, however, and the report resolved nothing. Three years later, with vaccination rates falling in the United States and anxiety among parents increasing rapidly, and after many calls from

physicians for clearer and more compelling guidance, the committee revisited the issue more directly.

Even at the height of the age of AIDS, when members of the
activist group ACT UP stormed St. Patrick's Cathedral, surrounded
the White House, shut down the New York Stock Exchange, and
handcuffed themselves to the Golden Gate Bridge, all to protest
the prohibitive cost of drug treatments and the seemingly endless time it took to test them, rancor between researchers and the
advocacy community was rare. The contempt AIDS activists felt
for federal officials—particularly for the Food and Drug Administration and its cumbersome regulations—was palpable. Even the
most strident among them however, seemed to regard physicians
as allies, not enemies.

Those days have ended, as the Institute of Medicine vaccine
committee came to learn. For years, the culprits most frequently
cited as the cause of autism had been the measles, mumps, and
rubella vaccine, as well as those that contained the preservative
thimerosal. Thimerosal was first added to vaccines in the 1930s in
order to make them safer. (Before that, vaccines were far more
likely to cause bacterial infections.) While descriptions of autistic
behavior have existed for centuries, the disease was only named in
1943—and its definition continues to evolve. Neurodevelopmen-
tal illnesses like autism have symptoms similar to those of mercury
poisoning, and there is mercury in thimerosal. What's more,
American children often receive a series of vaccinations when they
are about eighteen months old. That is a critical threshold in
human development, when a child often begins to form simple sentences and graduates from chewing or pawing toys to more
engaging and interactive forms of play. Some children don't make

that transition—and because they receive so many shots at the same time, many parents feared, naturally enough, that the inoculations must have been the cause.

Anguished parents, who had watched helplessly and in horror as their children descended into the disease's unending darkness, could hardly be faulted for making that connection and demanding an accounting. The Immunization Safety Review Committee was supposed to provide it, although its members represented an establishment trusted by few of those who cared most passionately about the issue. AIDS activism had its impact here too, because it changed American medicine for good: twenty-first-century patients no longer act as if their doctors are deities. People demand to know about the treatments they will receive, and patient groups often possess more knowledge than the government officials entrusted to make decisions about their lives. They have every right to insist on a role in treating the diseases that affect them.

The rise of such skepticism toward the scientific establishment (as well as the growing sense of anxiety about environmental threats to our physical health) has led millions to question the authority they once granted, by default, not only to their doctors, but also to organizations like the National Academy of Sciences. Faced with the medical world that introduced, approved, and relentlessly promoted Vioxx, a patient can hardly be blamed for wondering, "What do these people know that they are not telling me?" Uncertainty has always been a basic ingredient of scientific progress—at least until reason is eclipsed by fear. Unlike other commodities, the more accessible knowledge becomes, the more it increases in value. Many autism activists, however, sensed that federal health officials and researchers who work with them were guilty of avarice and conspiracy, or at least of laziness—guilty until proven innocent (and innocence is hard to prove). To use Fauci's formulation, when experts are tainted, where can you place your trust?

It was, in the highly emotional words of one vaccine activist who rejects the federal government's approach, a conspiracy among scientists to protect pharmaceutical companies at the expense of America's children. Because this is the age of denialism, evidence that *any* pharmaceutical company has engaged in venal behavior means that they *all* have. "When mothers and fathers take their healthy sons and daughters to pediatricians to get vaccinated and then witness them suffering vaccine reactions and regressing into chronic poor health within hours, days and weeks of getting sometimes five to ten vaccines on one day, they are not going to accept an illogical, unscientific explanation like 'it's all a coincidence,'" Barbara Loe Fisher has written. Fisher is the leader of the National Vaccine Information Center, the most influential of the many groups that oppose universal vaccination. (This sentiment, that children receive too many vaccinations when they are young, also draws frequently on anecdotal experience and the sort of conspiracy theories that are hallmarks of denialism: "No wonder our children are damaged and dealing w/ADHD, autism, diabetes, asthma, allergies, etc.," Fisher wrote. "Forty-eight doses of fourteen vaccines by age 6 is excessive and is only for the benefit of the drug companies who promote fear to fund their bottom line.")

The vaccine panel found itself at ground zero in this war against authority and scientific rigor. "It was the perfect storm," McCormick told me one day when we met in her office at Harvard. "Because all of a sudden we had the expansion of shots, the issue of the mercury in vaccines, and this rapid rise in the diagnosis of autism. Everyone put two and two together and came out with six." In a society where numeracy is rarely prized and subjective decisions often outweigh rational choices, it's not hard to understand at least some of the reasons why that happened: science works slowly and has yet to determine a cause for autism, which isn't even a single "disease," but rather a complex set of developmental disorders. In fact, it makes no more sense to talk about "curing" autism than it does to discuss a cure for cancer; "cancer" is an umbrella term for many diseases characterized by malignant growth. A successful treatment for leukemia won't stop the spread of melanoma. Effective treatments for autism will require a fuller understanding of how those developmental disorders differ—but they can differ widely. Autism spectrum disorders vary in severity from mild conditions like Asperger's syndrome to those characterized by sustained impairments in social interaction and communication abilities.

When "experts," often with degrees or licenses that seem impressive, suddenly emerge to tell heartbroken family members that there is a simple solution to their problems, who wouldn't want to believe it might be true? And with the help of the Internet, those experts are just a mouse click away. People often cling to their initial response when they discover something profoundly disturbing, even if more compelling evidence emerges. It's a form of denialism, but also a common human instinct. Not surprisingly, then, the vaccine panel was indeed "pummeled" during its deliberations. Attacks came by e-mail and over the telephone. One member was effectively forced to resign after he received an esca-

lating series of personal (and credible) threats that eventually became so worrisome that McCormick agreed to shift the venue of
the committee's final public meeting to a room where the members would be able to come and go in safety, interacting with the
audience behind a human moat of security guards. Like a jury
deciding the fate of a gangland leader, committee members were
encouraged to stay in a single hotel, discuss the location with no
one, and refrain from wandering about town on their own. Security was tightened; all this before the committee made its final
report. When it came time for the meeting, each member was
loaded onto a bus and driven directly to a garage beneath the main
building of the National Academy of Sciences. That way, they
could make it to the hearing room without having to run the
gauntlet of protestors.

THE REPORT, *Vaccines and Autism*, was issued in May 2004.
After an exhaustive analysis of the available data, and after review
by another independent panel, the committee concluded that
there was no evidence to suggest the existence of any relationship
between the two. "There really wasn't any doubt about the conclusions," McCormick said. "The data were clear." The Institute of
Medicine team attempted to review every important epidemiological study, whether published or not, involving hundreds of
thousands of children in several countries. They set out with a
clear goal: to discover what biological mechanisms involved in
immunizations might cause autism. Yet, no matter where they
looked or how they parsed the data, the central results never

varied: unvaccinated children developed autism at the same or higher rate as those who had been vaccinated. The panel reported accordingly: "The committee concludes that the body of epidemiological evidence favors rejection of a causal relationship between the MMR vaccine and autism."

The report also pointed out that the mercury contained within the preservative thimerosal, which had been used widely in vaccines for nearly seventy years, caused no apparent harm. Thimerosal had been a focus of special fury among anti-vaccine activists. By July 1999, however, two years before the IOM committee was convened, the preservative had been ordered removed from childhood vaccines as a precautionary measure. Vaccine manufacturers, under fierce public pressure, had agreed with the Centers for Disease Control and the American Academy of Pediatrics. "Parents should not worry about the safety of vaccines," the academy said at the time. "The current levels of thimerosal will not hurt children, but reducing those levels will make safe vaccines even safer. While our current immunization strategies are safe, we have an opportunity to increase the margin of safety." In other words, they decided it would be easier to get rid of the controversy than to explain it.

The decision, an attempt to placate parents, had no basis in scientific research, and set off a cascading wave of misunderstanding that persists to this day. Almost immediately, advocacy groups arose, filled with members who were convinced thimerosal had caused their children's autism. "It was a decision that was made very abruptly," McCormick, who had no role in making it, told me. "And with not very good communication between professionals and the public. Maybe they should have thought about what you might want to know to reassure people and that is a valid concern of parents. You know how this looks: 'Last year you told me this was safer than blazes, and this year you are taking it out of the drinking water. Hmm . . . how can I possibly trust a word you say?'" At the time, little was known about the toxicity of ethyl mercury, the chemical compound in thimerosal—so almost all toxicology data about mercury in vaccines was inferred from research into a related molecule, methyl mercury, which is found in fish that we eat and is used heavily in industry. While everyone has tiny amounts of methyl mercury in their bodies, the less the better, particularly because it can take months to be eliminated from our tissues.

The IOM examined the hypothesis that vaccinated children would develop a particular type of autism, caused by mercury poisoning. Presumably those children would develop symptoms at a different age than those who were not vaccinated. Yet in an analysis of tens of thousands of children, no statistical age difference was discovered. Furthermore, if, as so many parents and advocacy groups still believe, there is a link between thimerosal and autism, one would assume that the number of children diagnosed with the illness would have decreased rapidly after the middle of 2000, by which time the preservative had been removed from nearly every childhood vaccine. Researchers in Montreal had a unique opportunity to test this hypothesis because in the 1980s Canada began to phase out thimerosal slowly over a decade. As a result, Canadian infants born in the years between 1987 and 1998 could have received nearly any amount of thimerosal in vaccines, from none to 200 micrograms, the latter being nearly the maximum daily dose that had been permitted in the United States. The

Canadian team was able to study discrete groups and found that autism was most prevalent among children who received vaccines that contained no mercury.

Epidemiologists in Finland pored over the medical records of more than two million children, also finding no evidence that the vaccine caused autism. In addition, several countries removed thimerosal from vaccines before the United States. Studies in virtually all of them—Denmark, Canada, Sweden, and the United Kingdom—found that the number of children diagnosed with autism continued to rise throughout the 1990s, after thimerosal had been removed. All told, ten separate studies failed to find a link between MMR and autism; six other groups failed to find a link between thimerosal and autism. It is impossible to prove a negative—that a relationship between thimerosal and autism does not exist. While data can't prove that, it has failed to find any connection between them in any significant study. Because of the strength, consistency, and reproducibility of the research, the notion that MMR or thimerosal causes autism no longer seemed to public health officials like a scientific controversy.

The panel attempted to be definitive: "The committee also concludes that the body of epidemiological evidence favors rejection of a causal relationship between thimerosal-containing vaccines and autism. The committee further finds that potential biological mechanisms for vaccine-induced autism that have been generated to date are theoretical only. The committee does not recommend a policy review of the current schedule and recommendations for the administration of either the MMR vaccine or thimerosal-containing vaccines." The report suggested that people forget about thimerosal (which by then remained only in certain flu vaccines) and the controversy behind it. After all the research, thimerosal may be the only substance we might say with some certainty *doesn't* cause autism; many public health officials have argued that it would make better sense to spend the energy and money searching for a more likely cause.

It didn't take long for the findings, and the finality with which they were delivered, to generate reactions. Boyd Haley, professor of chemistry at the University of Kentucky and a witness who has often testified about his beliefs that mercury in thimerosal caused autism, was "amazed and astounded" that the IOM would conclude otherwise. Haley is also a prominent "amalgam protestor," convinced that trace amounts of mercury in dental fillings cause Alzheimer's disease, although no data exists to support that view. "The dismissal of thimerosal as causal to autism is outrageous!" he said. "It reflects a level of ignorance that is unacceptable for a scientific review committee. This is disgraceful and puts into question the very credibility of every oversight government authority in the United States. Exposure to thimerosal (mercury) causes a biochemical train wreck. I'm flabbergasted."

His was far from a lonely voice of outrage. The day the 2004 report was released, Indiana representative Dan Burton erupted. "This research does a disservice to the American people," he stated. Burton had long been a vociferous critic of the public health establishment and was well known for doubting many of the tenets of conventional medicine. "My only grandson became autistic right before my eyes—shortly after receiving his federally recommended and state-mandated vaccines," he said in 2002. In an October 25, 2000, letter to the Department of Health and Human Services, acting in his role as chairman of the House

Committee on Government Reform, Burton asked the agency's director to force the Food and Drug Administration to recall all vaccines containing thimerosal. "We all know and accept that mercury is a neurotoxin, and yet the FDA has failed to recall the 50 vaccines that contain Thimerosal," Burton wrote, adding, "Every day that mercury-containing vaccines remain on the market is another day HHS is putting 8,000 children at risk." (The letter was sent more than a year after voluntary withdrawal turned exposure to thimerosal, which remained in only some flu vaccines, into a rarity.)

It would be easy enough, and to many people comforting, to dismiss Burton as a fringe figure—his active indifference to scientific achievement and his opinions on health matters are highly publicized and widely considered ludicrous. If he is a fringe figure, however, he has unique power and many followers. The controversy not only continued, but intensified. Politicians have not shied away from using thimerosal as a public relations tool. On September 28, 2004, Arnold Schwarzenegger, governor of California, banned thimerosal-containing vaccines from his state for children and pregnant women. Other states soon did so as well.

No prominent American has spoken with more conviction about the putative dangers of vaccines or their relationship to autism than Robert F. Kennedy Jr. To him, the IOM report proved only that "the CDC paid the Institute of Medicine to conduct a new study to whitewash the risks of thimerosal," he wrote in 2005, "ordering researchers to 'rule out' the chemical's link to autism." That year, Kennedy, whose environmental work for the Hudson Riverkeeper organization has often been praised, published an article in *Rolling Stone* (and on the Internet at Salon.com) called

"Deadly Immunity." It was, he wrote, "the story of how government health agencies colluded with Big Pharma to hide the risks of thimerosal from the public . . . a chilling case study of institutional arrogance, power and greed. . . . I was drawn into the controversy only reluctantly. As an attorney and environmentalist who has spent years working on issues of mercury toxicity, I frequently met mothers of autistic children who were absolutely convinced that their kids had been injured by vaccines." He went on to say that he was skeptical until he read the scientific studies and looked at the data.

"Deadly Immunity" was a landmark in the history of science journalism, combining Kennedy's celebrity star power with a stinging assault on reason and scientific fact. The piece was riddled with inaccuracies, filled with presumptions for which there was no supporting data, and knit together by an almost unimaginable series of misconceptions. Kennedy largely framed his piece around quotes taken from the transcripts of a scientific meeting where members of the Immunization Safety Review Committee had gathered to plan their work. Those quotes appeared particularly to damn Dr. McCormick, the committee leader. According to Kennedy's article, McCormick told her fellow researchers when they first met in January 2001 that the CDC "wants us to declare, well, that these things are pretty safe," and "we are not ever going to come down that [autism] is a true side effect" of thimerosal exposure. In other words, before the committee even began its work, Kennedy asserts, McCormick had closed her mind to the possibility of a connection between thimerosal and autism. It was exactly the kind of conspiracy people concerned about the effects of the MMR vaccine had feared.

The transcripts tell a starkly different story. It's never hard to build a case with a partial quote; denialists do it every day, relying on fragmentary evidence and facts taken out of context. Here is what McCormick said: "I took away [from the previous day's discussion] actually an issue that we may have to confront, and that is actually the definition of what we mean by safety. It is safety on a population basis, but it is also safety for the individual child. I am wondering, if we take this dual perspective, we may address more of the parent concerns, perhaps developing a better message if we think about what comes down the stream as opposed to CDC, which wants us to declare, well, these things are pretty safe on a population basis."

As Harvey Feinberg, the former dean of the Harvard School of Public Health who is now head of the Institute of Medicine, pointed out at the time, the full quote was part of a discussion that focused on two issues: the need for parents to learn whether a vaccine was safe for an individual child who might be sick, and the public health community's right to know if vaccines pose risks to a larger population. In fact, McCormick proposed that the committee consider addressing the parental concerns about the health of an individual child—in addition to the CDC's concern about population-wide effects. McCormick's approach, her intentions, and her *words* were the opposite of what Kennedy had implied. But Kennedy was just getting warmed up.

"The CDC and IOM base their defense of thimerosal on these flimsy studies, their own formidable reputations, and their faith that journalists won't take the time to critically read the science," Kennedy wrote in 2007 in the Huffington Post, which has emerged as the most prominent online home for cranks of all kinds, particularly people who find scientific research too heavily burdened by facts. "The bureaucrats are simultaneously using their influence, energies and clout to derail, defund and suppress any scientific study that may verify the link between thimerosal and brain disorders. . . . The federal agencies have refused to release the massive public health information accumulated in their Vaccine Safety Database apparently to keep independent scientists from reviewing evidence that could prove the link. They are also muzzling or blackballing scientists who want to conduct such studies."

Kennedy has never explained why he thinks the public health leadership of the United States (not to mention its pediatricians) would wish to "poison an entire generation of American children." He simply wrote that "if, as the evidence suggests," they had, "their actions arguably constitute one of the biggest scandals in the annals of American medicine." In his *Rolling Stone* article, Kennedy ignored the scores of other published reports, few of which were carried out by federal scientists, so that he could focus on the 2004 study produced by the Institute of Medicine, which he attacked mercilessly. Kennedy wrote that vaccines exposed infants to 187 times the daily limit of ethyl mercury determined by the Environmental Protection Agency to be safe. If true, they would all have died at once. *Rolling Stone* soon printed a correction—and then later corrected that correction. It is impossible to live on the earth and avoid exposure to mercury, but that amount would kill a grown man.

The actual figure was 187 micrograms, which is 40 percent higher than the levels recommended for methyl mercury by the EPA, a tiny fraction of the figure cited in Kennedy's paper. Throughout the piece, Kennedy confused and conflated ethyl and methyl mercury, quoting as knowledgeable authorities the father-and-son team of Mark and David Geier, who have testified as expert witnesses in vaccine suits more than one hundred times—and who have been reprimanded repeatedly by judges who have dismissed them as unqualified to speak on the subject. (The father has an MD degree; David Geier holds an undergraduate degree in biology.) Their testimony has been tossed out of court on many occasions. One judge called Dr. Geier "intellectually dishonest," and another referred to him as "a professional witness in areas for which he has no training, expertise, and experience."

It is important to note that methyl mercury, the compound that is so dangerous when contained in fish and the product of industrial pollution, is *not* the mercury found in vaccines. The two forms differ by just one carbon molecule, which may seem insignificant. But as Paul A. Offit has pointed out in his indispensable book, *Autism's False Prophets*, a single molecule can mean the difference between life and death. "An analogy can be made between ethyl alcohol, contained in wine and beer, and methyl alcohol, contained in wood alcohol," Offit wrote. "Wine and beer can cause headaches and hangovers; wood alcohol causes blindness."

Kennedy saw conspiracy everywhere he looked. He has attacked Offit himself, who along with colleagues invented a vaccine to combat rotavirus, which is responsible for killing two thousand children in the developing world every day. Those children typically die of diarrhea, and in June 2009 the World Health Organization recommended that the vaccine be made part of "all national immunization programs." Kennedy, however, has referred to Offit, who is chief of infectious diseases at the Children's Hospital of Pennsylvania, as "Dr. Proffit" and as a "biostitute" because he was paid for his research and received royalties from the sale of his invention. Offit, outspoken and unremitting in his support of vaccines, has become a figure of hatred to the many vaccine denialists and conspiracy theorists. He has been threatened with violence so often that congressional aides once warned him not to mention the names of his children in public. For several years, armed guards have followed him to meetings of federal health advisory committees (where he has been called a terrorist), and employees in the mail room at Children's Hospital routinely check packages or letters addressed to him that look suspicious and might contain bombs.

Meanwhile, data becomes increasingly informative, particularly with regard to the difference between the effects of ethyl and methyl mercury. One of the biggest concerns researchers have always had about mercury was how long it took to be eliminated from a child's body. In 2008, a team of scientists at the Ricardo Gutierrez Children's Hospital in Buenos Aires published a report that examined the issue in detail. Kennedy had written that "truckloads of studies have shown that mercury tends to accumulate in the brains of primates and other animals after they are injected with vaccines—and that the developing brains of infants are particularly susceptible." It turns out that mercury in vaccines can be tolerated in far larger doses than was previously understood. More than two hundred children were studied after receiving vaccines that contained ethyl mercury, which is still used routinely in Argentina. The children excreted half the mercury within four days, and their levels returned to normal eleven days after vaccination. In contrast, it takes roughly seventy days for the body to flush half of a dose of methyl mercury.

No scientific report has seemed able to temper Kennedy's ardor or that of people like the actress Jenny McCarthy and her boyfriend, Jim Carrey, who have become America's marquee vaccine protestors. Like Kennedy, McCarthy knows how to wield her celebrity; in the fall of 2007 she appeared on Oprah Winfrey's television show, unleashing what she referred to as her "mommy instinct" in search of the cause of autism. "What number will it take for people just to start listening to what the mothers of children with autism have been saying for years—which is we vaccinated our babies and something happened. That's it," she said. When confronted with data from the Centers for Disease Control that seemed to provide scientific refutation of her claims, McCarthy responded, "My science is named Evan [her son] and he's at home. That's my science." McCarthy says that she "fixed" Evan by changing his diet, and recommends that other "warrior moms" do the same. She is fond of saying that she acquired her knowledge of vaccinations and their risks at "the University of Google."

Like Kennedy, McCarthy and Carrey contend that the federal government and pharmaceutical companies have conspired to keep the evidence that thimerosal-containing vaccines cause autism a secret. "In this growing crisis," Carrey wrote in the Huffington Post in April 2009, "we cannot afford to blindly trumpet the agenda of the CDC, the American Academy of Pediatrics (AAP) or vaccine makers. Now more than ever, we must resist the urge to close this book before it's been written. The anecdotal evidence of millions of parents who've seen their totally normal kids regress into sickness and mental isolation after a trip to the pediatrician's office must be seriously considered."

He had a point: every parent with an autistic child has the right to demand that federal researchers seriously consider anecdotal evidence. Anything less would be disgraceful. And that is why it has been considered in dozens of studies over more than a decade. Continuing to encourage false hope in this way, however, is an approach that Kathleen Seidel, whose blog Neurodiversity is the most complete and accessible collection of useful information about autism, has described quite accurately as nonsense, a litigation-driven hypothesis that autism is a consequence of vaccine injury.

Conspiracy theories are like untreated wounds. They fester and deepen—and the autism-vaccine conspiracy is no exception. Within days of Carrey's article, thousands of people had responded with comments on the Huffington Post Web site. Most were positive. Barbara Loe Fisher of the National Vaccine Information Center refers to the Public Health Service's insistence that the benefits of vaccines outweigh their risks as the "great denial": "It is only after a quarter century of witnessing the Great Denial of vaccine risks," she wrote, "which has produced millions of vaccine damaged children flooding special education classrooms and doctors offices, that the magnitude of that unchecked power has been fully revealed." Clearly, she is right about the powerful strain of denialism that the struggle over vaccines has exposed. She has the denialists and realists confused, however. That is one of the problems with conspiracy theories. After enough distortions seep into conventional thought, "the facts" look as they would in a funhouse mirror. Just tune in to YouTube and check out Robert F. Kennedy Jr. on the subject of vaccines. In June 2008, at a rally on Capitol Hill to "Green Our Vaccines"—in other words, to make them environmentally safe—Kennedy delivered his most inflammatory speech on the subject, saying that the "thimerosal generation is the sickest generation in the history of this country." It is not clear how he arrived at that conclusion, since life expectancy for newborns in the United States has increased dramatically over the past seventy years, from 57.1 for babies born in 1929 to 77.8 for babies born in 2004.

The change has been significant even during the past fifteen years—when Kennedy argues children have suffered the most. The trend is the same with regard to DALYs, or disability-adjusted life years, which measure healthy life expectancy—the number of years a child is likely to live without losing time to disability and sickness. In addition, during the period between 1990 and 2004, Kennedy's key danger years, childhood cancer death rates fell sharply (among both sexes, all ethnic groups except American Indians, and in every census region of the United States). It would be difficult to argue that any generation of children in the history of the world has ever been as healthy as the "thimerosal generation."

Yet, that sunny day in June, more than a thousand activists, most from groups like Talk About Curing Autism (TACA), Generation Rescue, Healing Every Autistic Life, Moms Against Mercury, and Safe Minds, a nonprofit organization that falsely characterizes autism as "a novel form of mercury poisoning," listened as Kennedy described the vaccination polices of the United States government as the "worst crime since the cover-up of the Iraq war," and added that the "the treatment of these children and the cost to our society" would far exceed the cost of the war itself.

DOES KNOWLEDGE SIMPLY disappear over time? After centuries of scientific progress have we not constructed pyramids of information solid enough to withstand periodic waves of doubt and anxiety? Human history has repeatedly suggested that the answer is no, but it isn't ignorance that makes people run from the past or shun the future. It's fear.

In 1421, China was far ahead of the rest of the world in sophistication, in learning, and particularly in scientific knowledge. It was the least ignorant society on earth. Then the newly completed Forbidden City was struck by a lightning bolt just as it opened, and the emperor reacted with horror. He interpreted the lightning as a sign from the gods that the people of the Middle Kingdom had become too dependent on technology—and were not paying enough attention to tradition or to the deities. So, as Gavin Menzies describes in his book *1421: The Year China Discovered America*, the Chinese burned every library, dismantled their fleets, stopped exploring the globe, and essentially shut themselves off from the outside world. The result? A downward spiral that lasted for five centuries. Japan, too, recoiled at progress by giving up the gun in the seventeenth century. Until they did, they were better at making steel than any Western country, and their weapons were more accurate, too. The samurai despised firearms. To them, they were nothing more than killing machines with the potential to destroy an enduring way of life. When somebody pulls a gun, it no longer matters how honorable you are or how many years you have trained with a sword. Guns put the social order in jeopardy, so Japan banned and eventually melted them all. The gears of social mobility were jammed into reverse. Again, it took centuries for Japan to regain its technological supremacy.

The history of vaccines, and particularly of smallpox, is filled with similar stories of fear at war with progress. According to Voltaire, the ancient Chinese inhaled dried powder of smallpox crusts through the nose in a manner similar to taking snuff. Thomas Jefferson's children were vaccinated by a slave who learned how to do it as a traditional aspect of African medicine. "We keep forgetting this stuff over and again," Juan Enriquez told me. Enriquez, who founded the Life Sciences Project at the Harvard Business School, is one of America's most insightful genomic entrepreneurs, and has spent many years studying the unusual ebb and flow of knowledge. "People become scared of change," he said. "They get scared of technology. Something bad happens and they don't know how to react: it happened with the emperor in China, and with smallpox, and it has happened with autism, too. People want to blame something they can't understand. So they blame technology. And we never stop forgetting how often this happens. We think of technology and the future as linear. It so clearly is not."

The controversy surrounding the MMR vaccine and autism is far from the first time the world has recoiled from vaccination, with at least some people convinced it does more harm than good. Smallpox arrived in Boston in 1721, carried by passengers on a ship from the Caribbean. It was the second coming of the epidemic to American soil—the first had landed more than one hundred years earlier. This outbreak was more severe, though, and by the time it had run its course, half of the ten thousand residents of Boston had fallen ill, and more than a thousand had died.

Cotton Mather, the fiery, brilliant, and unpleasantly self-righteous preacher, had heard about vaccination several years earlier from an African slave. He then read about the practice in a British scientific journal and became convinced it could provide the answer to the plague that threatened the city (and the entire New World). Mather attempted to interest the town's residents in what he acknowledged was the genuinely risky "Practice of conveying and suffering the Small-pox by Inoculation," a practice "never used . . . in our Nation."

There were few takers. Instead, the majority of the population was awed by the ability of smallpox to wipe out entire nations and wondered whether it was not simply a judgment from God, rather than a disease one could defeat with medicine. Most people condemned inoculation. Perhaps the answer was to turn inward, to pray more fervently. Mather screamed from the pulpit (joined by several others, including his father, Increase Mather—they came to be known as the Inoculation Ministers). They faced opposition from the nation's first powerful newspaper, known at the time as the *New England Courant* (eventually to become the *Hartford Courant*), which was published by Benjamin Franklin's brother James—and not just from him. "Cotton Mather, you dog, damn you! I'll inoculate you with this; with a pox to you," said a note that was attached to a bomb lobbed into Mather's house. All because he argued for the adoption of the most important public health measure in the history of colonial America.

Ben Franklin himself opposed the idea of the inoculation—called variolation, in which healthy people would have pus from the scabs of smallpox victims rubbed onto their skin. This usually produced a much milder form of smallpox, although a small percentage of the people vaccinated in this way died as a result. When the final tally was made, however, the salutary effects of vaccinations were impossible to deny. Of the 240 people inoculated during the epidemic in Boston, six died, one in forty. Among the rest of the population the mortality rate was one in six. Even those made sick by the vaccine tended to become less seriously ill than those who acquired the infection in the usual way.

Years later, Franklin's son died of smallpox, after which he became an ardent supporter of vaccination. He even made a special appeal to parents who might be afraid of the consequences. "In 1736 I lost one of my sons, a fine boy of four years old, by the smallpox taken in the common way," he wrote in his autobiography. "I long regretted bitterly and still regret that I had not given it to him by inoculation. This I mention for the sake of the parents who omit that operation, on the supposition that they should never forgive themselves if a child died under it; my example showing that the regret may be the same either way, and that, therefore, the safer should be chosen." George Washington initially hesitated to vaccinate his Continental Army troops during a smallpox outbreak, writing that "should We inoculate generally, the Enemy, knowing it, will certainly take Advantage of our Situation." By 1777, however, he ordered mandatory vaccination for every soldier.

Vaccines work primarily by stimulating the immune system to produce a defensive response; there is a small risk that the response won't be good enough and the vaccine will cause the disease it has been designed to prevent. Unless you compare those risks with the alternative—that is, of not having the vaccine at all—there is no way to properly judge any vaccine's value to society. It has been three hundred years since Tony Blair's distant predecessor, Robert Walpole, Britain's first prime minister, purchased vaccines for the children of King George I. "Here is the tradition that Tony Blair is fighting against," Enriquez said. We were sitting in the parlor of his house outside Boston—which had once served as a stop along the Underground Railroad. Enriquez is a collector of medical records, scientific charts, and many types of maps. On the table before us he had spread his "smallpox collection," letters relating to the history of the disease. He produced a yellowed parchment that looked more like an eighteenth-century proclamation than a bill. "This is from a guy called Charles Maitland, a surgeon," he said. "It's a bill of sale for vaccinating the king's children in 1724, signed by Walpole. So here is the British royalty in 1724 understanding that it is really important to vaccinate your children. It is just amazing to me that you can take this . . . and move to where we are today."

At that, Enriquez stood up to fetch another armful of documents. "These are Jenner's notebooks on vaccination," he explained. Edward Jenner is generally credited with having invented the smallpox vaccine, after noting that milkmaids rarely got the disease. He theorized, correctly as it would turn out, that the pus in the blisters that milkmaids received from cowpox (a disease similar to smallpox, but much less virulent) protected them from smallpox. "Here are the letters to the public health department in 1804 donating the skins that Jenner had tested. And here are the enclosed tests." It was all there: the data, the evidence, sitting irrefutably three centuries later on a table in a suburb of Boston. "We have this science," Enriquez said, pointing respectfully at the notebooks. "Look at the data, it's so clear. Here is the result of the first twenty tests. It's a pile of stuff telling us something the king and his circle knew instinctively in 1724: vaccinations are essential."

Instincts evolve. We have become inherently suspicious of science, so when a drug company or a researcher does something wrong, fails to show data that could be harmful, for example, or when there's an issue having to do with the safety of a particular product, it feeds into the underlying suspicion and permits people to say, "Ah! All of science is bad." When people encounter something that isn't immediately explicable—autism, for example—it plays into this sense of doubt, and even when the scientific evidence is overwhelming people don't always believe it. It is a climate that has created people like Tony Blair's wife, Cherie, who has long been known for her skepticism toward many aspects of conventional medicine. She recently acknowledged that while she had been highly suspicious, she did eventually vaccinate her son Leo. Blair had been influenced by, among others, a half sister who had criticized the bonus payments doctors in England received for administering MMR shots, and publicly declared that she would never vaccinate her own daughter. "A number of people around me, whose views I respected, were vociferously against all forms of vaccination," Blair said in 2008. "Over the years I had listened to their side of the argument and, it's fair to say, I was in two minds."

INCREASINGLY STYMIED in their quest to blame autism on the mercury contained in thimerosal, or on vaccines in general, unconvinced by mounting evidence that genetics and the environment play significant roles in the development of autism, activists began to hunt for a new approach to bolster their vaccine theory. On March 6, 2008, they found what they were looking for. On that day, the parents of Hannah Poling, a nine-year-old girl from Athens, Georgia, held a press conference to announce that the Department of Health and Human Services had issued what Jon Poling, the girl's father, referred to as a ruling that may "signify a landmark decision with children developing autism following vaccines." He said he made the public statement to provide "hope and awareness to other families." His lawyer, Clifford Shoemaker, who has turned the vaccine plaintiff industry into a lucrative career, was by his side. They both nodded when a reporter asked whether the case was the first in which "the court has conceded that vaccines can cause autism."

Nearly five thousand families have similar suits before the court, so any judgment that applied broadly would have had profound implications. Paul Offit has described Hannah Poling's case in the *New England Journal of Medicine*. When she was nineteen months old, Hannah Poling received "five vaccines—diphtheria–tetanus–acellular pertussis, *Haemophilus influenzae* type b (Hib), measles–mumps–rubella (MMR), varicella, and inactivated polio. At the time, Hannah was interactive, playful and communicative girl. Two days later she was lethargic, irritable, and febrile. Ten days after vaccination, she developed a rash consistent with vaccine-induced varicella. Months later, with delays in neurologic and psychological development, Hannah was diagnosed with encephalopathy," a syndrome characterized by altered brain function, "caused by a mitochondrial enzyme deficit," which left her weak and often confused. Hannah's symptoms, Offit wrote, "included problems with language, communication, and behavior—all features of autism spectrum disorder. Although it is not unusual for children with cellular disorders like mitochondrial enzyme deficiencies to develop neurologic symptoms between their first and second years of life, Hannah's parents believed that vaccines had triggered her encephalopathy. They sued the Department of Health and Human Services for compensation under the National Vaccine Injury Compensation Program." When a federal court awarded damages to their daughter, the vaccine scare was ignited anew.

The vaccine court is unlike any other in the United States, however, and the suit can only be examined in the context of the way it works. "In the late 1970s and early 1980s, American lawyers successfully sued pharmaceutical companies, claiming that vaccines for pertussis caused a variety of illnesses, including unexplained coma, sudden infant death syndrome, Reye's syndrome, mental retardation, and epilepsy. By 1986 all but one manufacturer of the diphtheria–tetanus–pertussis vaccine had abandoned the American market." The risk of lawsuits became far greater than the potential for profits. "The federal government," Offit continued, "increasingly concerned that no company would be willing to manufacture essential vaccines, passed the National Childhood Vaccine Injury Act, which included the creation of the VICP.

"Funded by a federal excise tax on each dose of vaccine, the VICP compiled a list of compensable injuries. If scientific studies supported the notion that vaccines caused an adverse event—such as thrombocytopenia" (the dangerous depletion of platelets) "after receipt of measles vaccine, or paralysis following an oral polio vaccine—children and their families were compensated," and usually quite generously. The average family received nearly $1 million. Health officials developed a table of injuries that would apply, and the number of lawsuits against vaccine manufacturers fell dramatically. One of those "table injuries," encephalopathy following a measles vaccine, is the cellular disorder for which Hannah Poling was awarded money.

Because some of the court records are sealed, it is impossible to know with certainty how the decision was made. Nonetheless, Hannah's father, Jon Poling, and Shoemaker, the lawyer, were wrong: neither the court nor the federal government conceded that vaccines cause autism. They conceded that in that specific case vaccines may have been responsible for exacerbating a condition with symptoms that are similar to those of many autistic disorders. Hannah had a mitochondrial enzyme deficiency, which consisted of a metabolic disorder called encephalopathy. (Mitochondria are responsible for producing 90 percent of the energy we need to stay alive. Many defects can prevent them from functioning properly or at all—and without enough energy, cells, like any other factory, stop working.)

The court was compelled to address a difficult question: could the fever that Hannah developed following those vaccines (one of which was a measles vaccine) worsen her encephalopathy? "This is a particularly complex issue," Anthony S. Fauci, the director of the National Institute of Allergy and Infectious Diseases, told me. "And it's even more complicated by the fact that it is likely that there are a minute fraction of kids who have an underlying defect—clearly, mitochondrial defects are the one that stands out—where at a certain time of their life, when they get confronted with an influenza, with an [environmental toxin] or a vaccine, it's going to accelerate what already was going to happen anyway. Then some people will say that every autistic kid became autistic because of a vaccine. And we know absolutely that that's not the case."

In early 2009, the vaccine court agreed. After ten years of bitter scientific and legal battles, the court rejected any relationship between autism and vaccines. "Petitioners' theories of causation were speculative and unpersuasive," wrote Special Master Denise Vowell in the case of *Colten Snyder v. HHS.* "To conclude that Colten's condition was the result of his MMR vaccine, an objective observer would have to emulate Lewis Carroll's White Queen and be able to believe six impossible (or at least highly improbable) things before breakfast."

It was a decisive victory for science—but denialists have no interest in scientific decisions. "The vaccine court was wrong," Jay Gordon wrote on the Huffington Post the next day about the nearly thousand pages' worth of rulings in three cases. Gordon, or Dr. Jay as he calls himself, is read by many people and serves as, among other things, Jenny McCarthy's pediatrician. He added, "Let me state very simply, vaccines can cause autism. . . . The proof is not there yet. It will be found." His judgment on the court and the ruling was equally certain. "They were disdainful and unscientific in their approach and did not gather the needed evidence. . . . Vaccines as they are now manufactured and administered trigger autism in susceptible children." (While rejecting scientific evidence themselves, denialists often explain their position by citing the "unscientific" approach of opponents.)

Nobody can blame any parent of an autistic child for skepticism, confusion, or anger. But when confronted with statements like Gordon's, an honest researcher could only shake his head in wonder. "If it wasn't for suspicion about science, I think the lack of that computing rationally would dawn on people," Fauci said. "This is ridiculous! You have all these studies, you have this, you have that, it doesn't matter. And it's a combination of the underlying mistrust and the fact that people with autistic children see them suffering horribly and they need an explanation. What's the explanation? Well, we don't have one. Instead we have fear, and fear is what propels this movement."

Because Hannah Poling's encephalopathy included features of autism, it raised questions about how autism is defined and why case numbers have expanded so rapidly. Many reasons seem to have converged: new and broader diagnoses for Medicaid and insurance, increased awareness, and more access to services are all among them. The federal Individuals with Disabilities Education Act, passed in 1991, guarantees appropriate public education for children with autism. Soon after the law was enacted, schools began reporting an increase in the number of autistic students. That is one reason the act was written, but to receive special education a child must have a specific diagnosis. In the past, parents without a clear sense of what was wrong with their children were often vague. They had no choice, because states presented few options. Autism now appears on application forms for state aid, and it is one of the categories parents can consider when seeking special educational support.

Once again, the confusion surrounding the Poling decision was magnified by the Internet, causing a new wave of fear, anxiety, and, naturally, reluctance on the part of many parents to vaccinate their children. It also played into the anti-vaccine movement's newest catch-phrase: "Too Many Too Soon." The phrase implies that there are all sorts of toxins in our vaccines and that by receiving so many shots at such an early age, children's immune systems are easily overcome. Now that thimerosal has proven ineffective as a propaganda weapon, the anti-vaccine forces have begun more frequently to invoke "Too Many Too Soon" and "Green Our Vaccines" as standard tropes.

As David Gorski, a surgeon who teaches at the Wayne State University School of Medicine, has pointed out in the blog Science-Based Medicine, they are brilliant slogans. Who doesn't want a green vaccine? Jenny McCarthy likes to say she is not opposed to vaccines, just to the poisons they contain. "This is not an anti-vaccine rally or an anti-vaccine group," she said at the Green Our Vaccines Rally in Washington in 2008. "We are an intelligent group of parents that acknowledges that vaccines have saved many lives, but what we're saying is that the number of vaccines given [needs to be reduced] and the ingredients, like the freakin' mercury, the ether, the aluminum, the antifreeze need to be removed immediately after we've seen the devastating effect on our children.

"I know they are blaming me for the measles outbreak and it cracks me up," she continued. "What other consumer business blames the consumer for not buying their product? It's a shit product. Fix your product. That's what they are trying to do: scare people into vaccinating again and point the blame." In April 2009, McCarthy and Carrey began a national tour to promote her third book on autism, *Healing and Preventing Autism: A Complete Guide*, written with McCarthy's "biomedical" adviser, Dr. Jerry Kartzinel.

Larry King has always been among McCarthy's most enthusiastic enablers, and he was one of the couple's first stops on the tour. King asked McCarthy why she thought pharmaceutical companies would expose children to so many shots and such danger. "Greed," she said, making note of the increased number and type of vaccinations in the past twenty years. King asked: because they all want to make money? "Of course," McCarthy replied.

That week, in an interview with *Time* magazine, McCarthy actually said she thought that old diseases would have to return—as in fact they have. "I do believe sadly it's going to take some diseases coming back to realize that we need to change and develop vaccines that are safe. If the vaccine companies are not listening to us, it's their fucking fault that the diseases are coming back. They're making a product that's shit. If you give us a safe vaccine, we'll use it. It shouldn't be polio versus autism." As she has repeatedly, McCarthy cited the number of vaccines children receive as a large part of the problem.

Nobody can dispute that the number of vaccines has multiplied enormously. One hundred years ago children received a single vaccine: smallpox. By 1962 that number had grown to five (diphtheria, pertussis, tetanus, polio, and the MMR). Today children are vaccinated for eleven diseases, and by their second birthday that can entail as many as twenty separate shots. Yet because molecular biology has made it possible to create vaccines with fewer antigens, children are exposed to far less of a burden than children in past generations had been. The smallpox vaccine, for example, contained two hundred proteins—all separate molecules. Together, the eleven vaccines that children routinely receive today contain fewer than 130. The bacteria that live on the nose of a newborn child or the surface of their throat number in the trillions. "Those bacteria have between 2,000 and 6,000 immunological components and consequently our body makes *grams* of antibody to combat these bacteria. . . . The number of immunological challenges contained in vaccines is not figuratively, it is *literally*, a drop in the ocean of what you encounter every day," Paul Offit has written.

The Poling decision opened the door to a new round of debates, and one of the first to rush through it was Bernadine Healy, a former director of the National Institutes of Health. She stunned parents and scientists alike in 2008 by saying that while "vaccines are safe, there may be this susceptible group. I think that public health officials have been too quick to dismiss the hypothesis as irrational." That year, the anti-vaccine blog the Age of Autism, which describes itself as the "daily web newspaper of the autism epidemic," made Healy its person of the year. She often tries to present herself as a calm voice of reason caught between two equally emotional camps. Following Jenny McCarthy's appearance on *Larry King Live* in 2009, however, Healy ventured even deeper into denialism. "The work is long overdue; shockingly, so is a study comparing groups of vaccinated and unvaccinated children," she wrote on the blog she maintains for *U.S. News and World Report*. Actually, those studies, involving more than a million children, have been carried out in several countries and for many years.

"How many studies are enough?" Fauci asked. "The Institute of Medicine did all it could do. But the real problem here is that no politician can afford to appear as if he or she is brushing off the agonizing concerns of a parent with an autistic child. So none would say what needed to be said. I would have liked to have seen some political leaders say, Folks, come on," Fauci said. "We're going to put fifty million more dollars into autism research and look at what the real causes might be. But it's dangerous politically to do that, because of the intensity of the people who believe that vaccines are the problem. When you have a suffering family, and an innocent kid who's got a life that is going to be difficult, it's very, very tough to go and make a pronouncement that this is not the case. Particularly because you will then have that rare event when a vaccine might precipitate an underlying genetic defect that might just as well have been set off by something else, like the flu. It's just dangerous politically for anybody to speak out. It's not very courageous not to do it, but when political people look at all the things they're going to step up to the plate on, that's not one of them."

IT'S HARD TO GET to Marie McCormick these days. You have to make it past two guards at the entrance of her building at the Harvard School of Public Health. There is another guard not far from her office. Most of the time, she lets assistants answer the phone, and she sends a lot of e-mail to the spam box, too. McCormick has been subpoenaed to appear in a vaccine compensation case—not an occurrence she envisioned when she took the position as head of the Institute of Medicine's Immunization Safety Review Committee. She remains cheerful and committed to her work, yet she and other public health officials have watched as a steady stream of troubling facts emerged over the past decade:

more parents are choosing to homeschool their children, and while that is usually for economic or religious reasons, surveys suggest that an increasing number of parents do it because vaccinations are not mandated for most children schooled at home.

In Ohio, for example, the number of religious and philosophical exemptions for vaccines nearly quadrupled between 1998 and 2008, from 335 to 1,186. (In most states, though, it is actually easier to get exemptions from vaccine requirements than to have your child vaccinated.) Maine led the nation in immunizing children in 1996 with a compliance rate of 89 percent. By 2007, that figure had fallen to just 72.9 percent. The numbers vary widely even within states. On Vashon Island, an upscale suburb of Seattle, nearly 20 percent of parents have opted out of vaccination, and as the number of exemptions grows, infectious disease proliferates. In early 2009, after a lengthy investigation, the *Los Angeles Times* found that while the California state exemption rate was only 2 percent, the bulk of those exemptions were clustered in just a few school districts.

The trend lines are clear. In 1997, 4,318 California kindergartners entered school with vaccine exemptions. By 2008 the number—with fewer students enrolled—had more than doubled. At Ocean Charter School near Marina del Rey, 40 percent of kindergartners entering school in 2008 and 58 percent entering the previous year were exempted from vaccines, among the highest rates in the nation. These numbers are particularly discouraging because they threaten herd immunity. The term refers to the fact that if you are unvaccinated and everyone around you—the herd—is, there would be nobody who could conceivably spread the illness. The herd would protect you. When vaccination rates

drop significantly below 90 percent, however, herd immunity begins to disappear and infectious diseases can spread. Without herd
immunity, math takes over, and the results are easy to predict. The
winter after Wakefield's *Lancet* article appeared, for example, more
than one hundred children were hospitalized in Dublin, three of
whom died. Until that year, Irish public health officials regarded
measles as a disease that had essentially been eradicated.

Viruses are tenacious and they multiply rapidly; without vigi-
lance, past successes are quickly erased. By June 2008, as a result
of the declines in vaccine coverage, measles became endemic in the
United Kingdom fourteen years after it had practically vanished.
That means that for the foreseeable future the disease will recur
with predictable regularity. Since April 2008, two measles-related
deaths have been reported in Europe, both in children who, due to
congenital illnesses that affected their immune systems, were unable to receive the MMR vaccine. Children like that depend on
herd immunity for protection from an infectious disease like measles, as do children under the age of one, who normally are too
young to have received the vaccine. Unless the vast majority of
children are vaccinated, herd immunity disappears.

It is a remarkable fact that measles is on the rise in countries
like Britain and the United States even as death rates have plummeted over the past five years in the developing world. The United
States had more measles infections in the first half of 2008 than
in any comparable period for nearly a decade. There are regions
where half the children remain unvaccinated. Measles is particularly virulent and moves rapidly. "Somebody who has taken an
exemption from school laws, like a philosophical or religious exemption, is thirty-five times more likely to get measles . . . and

twenty-two times more likely to get whooping cough," said Dr. Lance Rodewald, director of the Centers for Disease Control's Immunization Services Division. Usually in the past, deaths began to occur when the caseload reached into the hundreds. Many public health officials feel that is now inevitable.

The threats are not theoretical—nor are they confined to measles. Early in 2009, five children fell ill after an outbreak of Hib influenza (*Haemophilus influenzae* type B) in Minnesota. One of the children died—the first Hib fatality in the state since 1991, the year after a vaccine became available. Until then, severe Hib infections would typically sicken twenty thousand children under the age of five every year in the United States, resulting in about a thousand deaths. Of the five Minnesota cases, three—including the child who died—were infants whose parents had refused to vaccinate them.

There has been a concerted international effort, led by the Bill and Melinda Gates Foundation and the World Health Organization, to eradicate polio. But success has been elusive, largely because of opposition to vaccines in northern Nigeria. In the summer of 2003, Muslim clerics banned polio vaccinations, claiming that the drugs were a Western plot to spread HIV and sterilize Muslim girls. People stopped taking the vaccine, and the results were disastrous: HIV rates didn't change but polio infections rose rapidly; some of those people traveled to the hajj in Mecca, infecting others. Soon there were cases in a dozen other countries. A world on the verge of eradicating polio would have to wait for another day. "Never forget that the longest flight any person can take anywhere in the world is shorter than the shortest incubation period of an infectious disease," McCormick said. "So you really have to understand that it is not only possible, but inevitable, that if people aren't vaccinated they will get sick."

Nearly every vaccine is under some form of attack. In 2007, the conservative legal group Judicial Watch announced that it had uncovered three deaths linked to the human papilloma virus (HPV) vaccine, which has been the subject of much debate around the country as states decide whether to require vaccinations before permitting girls to enter sixth grade. HPV is the most common sexually transmitted disease in the United States; at least half of all sexually active Americans become infected at some point in their lives. The virus is also the primary cause of cervical cancer, which kills nearly five thousand American women every year and hundreds of thousands more in the developing world.

The vaccine has proven exceptionally effective at preventing those infections. But Judicial Watch's report was picked up throughout the country (and the world) and often reported as if it were true. Each death was registered in the Vaccine Adverse Events Reporting System (VAERS), and each concluded that the girls died of heart attacks after receiving the HPV shots. When officials from the Centers for Disease Control talked to doctors and examined the dead girls' medical charts, however, they found that one death, a case of myocarditis, was caused by influenza virus. The other two were a result of blood clots in adolescents taking birth control pills, which are known to raise the risk of blood clots. "These deaths are tragic, but appear to have causes unrelated to vaccination," said Dr. John Iskander, co-director of the Immunization Safety Office at the CDC. Nonetheless, another front in the war against progress had been opened, and there have been other attempts to dismiss the importance of this new and effective cancer vaccine.

Perhaps that should surprise no one, since the fear of a common disease like autism will almost always outrank a fear about something like measles that people no longer take seriously. "Most parents I know will take measles over autism," J. B. Handley said not long ago when asked, for perhaps the thousandth time, why he persists in opposing the MMR vaccine. Handley is founder of Generation Rescue, an organization of parents who remain strongly committed to the idea that vaccines cause autism. Is that really a choice we are prepared to let him make? By choosing not to vaccinate their children, parents are not protecting them from autism—as so many epidemiological studies have demonstrated. They are simply putting their children—and the children of their neighbors—at greater risk of contracting diseases that could send them to the hospital or worse. How many American children will have to die in order to make the point that vaccinations are vital? How far will we descend into denialism before the fear of disease once again overshadows the fear of vaccines?

3

The Organic Fetish

It would be hard to argue with Dostoyevsky's assertion that a society's level of civilization can best be judged from its prisons. If you want to talk about aspirations, however, a supermarket might be a better place to look—and in America today that supermarket would have to be Whole Foods. It is no longer necessary to hunt for a small co-op if you want to buy organic peanut butter or grass-fed beef, of course. Thousands of stores offer organic and "natural" products, and sales are growing at several times the rate of more conventional fare, even in the midst of a recession. In 2008, that amounted to nearly $23 billion in the United States alone. Even places like Wal-Mart and Costco have heard the message; they have become the biggest organic distributors of all.

But as an emblem of the nation's growing fixation on all that is natural, nothing quite compares to Whole Foods.

To saunter down its aisles is to wade into a world of butter lettuce, chard, black radish, winter squash, and several types of arugula. The origin of nearly every product is on display, the better to assess its carbon footprint, the burden it places on the environment, and the likelihood that the food is fresh. Signs at the meat counter promise animals raised without hormone injections or antibiotics and nourished only by vegetables. Whole Foods adheres to the "Organic Rule," which, according to one of the store's many informational brochures, *Organics and you,* is principally about integrity.

Traditional grocery stores rarely made much of a fuss about their philosophies, but Whole Foods isn't just about food, it's about living a particular kind of life, an approach the company sums up nicely in the Whole Foods Credo: "Eat seasonally grown food, reduce the distances from farm to plate, shrink one's carbon footprint, promote sensitivity and a 'shared fate.'" (Eating locally produced food has become such a phenomenon that in 2007 the editors of the *New Oxford American Dictionary* selected "locavore" as their word of the year.) The credo is not unusual. Organic products nearly always come with claims of ethical superiority. Nature's Path, for instance, points out on its cereal boxes that traditional desires for profit and brand leadership would never "amount to a hill of beans if we don't choose sustainable, environmentally responsible processes that will leave the world better than we found it."

Calls for sustainability are no longer sales pitches or countercultural affectations; they have become a governing mantra of the progressive mind. "If you are concerned about your welfare, or your children's health, about the way animals are treated," Peter Melchett, the policy director of the British Soil Association, said, "or if you are concerned about the welfare of farmers or the future of this planet, you should buy organic food." Lord Melchett is one of the Western world's best-known and most outspoken believers that organic food, deployed properly, could feed humanity.

I enjoy shopping at Whole Foods. The products, while so costly that the chain's most frequently used nickname is "Whole Paycheck," are good and usually fresh. The colors are pleasing, the aisles roomy, and customers you find there seem more animated than at any other grocery store. I once stood at the entrance to the store at New York's Time Warner Center and asked fifty people why they were willing to spend extra cash—sometimes twice the cost of more traditional fare—to buy organic produce. Most said they thought organic food, which is grown without the use of synthetic pesticides or genetically engineered ingredients, tasted better, and many believed it would improve their health; still others hoped that a wider reliance on organic food might contribute to a better fate for the earth.

Americans seem to be thinking about their nutritional choices in a way that they never have before, and who can blame them? Our most significant health problems are diseases of indolence and obesity. We eat the wrong things. And we eat too much of them. America's top food group, measured by caloric consumption, is "sweets," and a dollar spent on junk food will buy many more calories than one spent on fruits or vegetables. Processed foods and mindless consumption have long been the twin pillars of our culinary experience.

"What kind of society kills itself by eating?" a woman asked me one day while shopping at Whole Foods. "We are not only ruining ourselves, we are ruining the land. How much of this planet do we have to plow under before we stop and realize we can live on the earth without destroying it? This is a moral issue as much as it is any other kind." That is a noble sentiment, expressed frequently within the organic movement, and it reflects a clear mission: to cultivate and eat what is wholesome and natural and to cleanse the world of its heavy reliance on synthetic, genetically manipulated, and chemically treated foods. While nowhere near as rabid on the subject as Europeans, who continue to confront transgenic crops with animosity and even violence, surveys demonstrate repeatedly that Americans consider organic foods far healthier than any product of biotechnology.

There's a lot to be said for buying locally grown produce: it can help sustain community farmers and focus attention on the quality of the environment. It tastes better, too. But is organic food healthier for you than food that contains genetically engineered ingredients or that has been harvested by robot-guided combines instead of human hands? Is it more likely to sustain the planet or the majority of its inhabitants? And are organic fertilizers and pesticides clearly a more virtuous and earth-friendly choice for the consumer than those made of synthetic chemicals? There are no short answers to those questions (at least none that are true). But there has certainly never been a study that would suggest the answer to any of them is a simple yes. There is no evidence, for example, that a single person has died or become seriously ill as a result of the accumulated residue of pesticides in their food. The same cannot be said of the toxins contained in "natural" food—as any number of salmonella outbreaks or raw milk poisonings in the United States continually demonstrate. In 2009, after salmonella and listeria contamination sent dozens of people to hospitals in six states, the Food and Drug Administration even warned Americans to avoid raw alfalfa sprouts—perhaps the signature food of a healthy, organic lifestyle.

"People buy many things when they buy organic food," Marion Nestle told me one day as we poked around one of Whole Food's newest and largest stores, the 69,000-square-foot colossus in the TriBeCa section of lower Manhattan. "One thing they believe they are buying is a nutritionally superior product. The entire organic industry is desperate to show that it has more nutrient content than conventional food. There may be more nutrients in some organic products, but so what? We get all the nutrients we need. In this country that is not our problem. Calories and diabetes and sugar. That's our problem."

Nestle, the author of *Food Politics* and *What to Eat,* and for many years a professor of nutrition at New York University, has been relentless in calling attention to false and confusing claims made by food companies (and permitted by the federal government). She is also a fan of organic food. "Buying what is grown with care is never wrong," she said. "But we have to pay attention to the differences between health claims and marketing ploys, and a lot of those labels"—she waved vaguely at the supermarket shelves around us—"purposefully confuse the two."

I had asked Nestle to join me in a quest to find one completely natural product for sale on the shelves of Whole Foods: something unaltered from what it would have been had we found it in its wild state. Americans often act as if the world around them was pristine until corporations began to defile it. Small-scale farming is an expression of that belief—an attempt to step away from the vast agricultural conglomerates that churn out Brussels sprouts and broccoli by the ton, as if they were auto parts or computer chips. There is, however, almost no such thing as natural food for sale in American grocery stores. Even spring water is processed (and, obviously, bottled). Salt is usually iodized. Fruit needs to be refrigerated or it will rot. Are organically grown carrots and celery, cut conveniently into snack-sized strips, wrapped in plastic and resting on a bed of Styrofoam, natural? "The first third of the store is fine," Nestle said, referring to the fresh vegetables, meat, and fish. Most had been refrigerated and trucked in from great distances—often at environmental costs that are difficult to calculate. But still, these foods contained no additives. Then she cast her gaze across the long rows of potato chips, artisan sugars, and high-end crackers. "That, they should get rid of," she said. "It's junk food."

Nestle reached for a box of organic instant oatmeal with Hemp Plus, which was fortified by omega-3 fatty acids. The consumer information on the back pointed out that the hemp contained neither marijuana nor any psychoactive drug. If the drug-free status of the cereal wasn't enough to entice consumers, the label also noted that both Thomas Jefferson and George Washington farmed it extensively.

"And of course they are pushing the omega-3s," Nestle said. "Omega-3 fatty acids are among the hottest ingredients right now." When eaten in trout, salmon, and other fish, omega-3s have been shown to lower the risk of heart attacks. But not all fatty acids are created equal; in cereal it is hard to know which omegas, if any, you are getting and whether they will be digested and absorbed properly by the cells they are supposed to protect. "Cereal is not a fish," she said. "And this is just another way to market a bunch of calories."

Organics still only account for a sliver of American food products, less than 5 percent, but that sliver is growing rapidly. Most of the crops in the United States, however, including 90 percent of the enormous soybean crop and more than three-quarters of the corn, are products of biotechnology. In 2008, 62.5 million hectares of genetically engineered food were planted in the U.S. In the rest of the world, the figure is growing by about 10 percent a year. To many inhabitants of the rich countries of the West, this suggests that we have become too reliant on technology for our food, that somehow the cold and soulless hand of science has been placed in nature's way. That's where the Whole Foods Credo comes in. "A shared fate" would require a sensible effort to keep destructive growth in check, and to find harmony in a world that is rapidly becoming depleted.

It would be hard to question the judgment of people who have no desire to eat hormone-infused meat or foods that have been processed and glued together by little more than a variety of sugars and fat. One has to wonder though, about that idea, so commonly espoused by environmental organizations like Greenpeace, Friends of the Earth, and the Organic Consumers Association, of a shared sense of fate. Exactly whose fate do these people think we are sharing? If it is the other billion or so residents of the rich world—the relatively few who can afford to shop at greenmarkets, eat tomatoes that still cling to the vine, and would rather dine at a restaurant that has been cited by the local health department for rodent infestations than at one that serves food trucked in from an industrial farm—then sure, our fate is shared. I have seen how most American chickens spend their lives, and nobody should help inflict that kind of misery on any living creature by buying battery-raised poultry or eggs.

It doesn't take a visionary, however, to understand that the other five billion or so residents of this world, more than half of whom live on less than two dollars a day, can't afford organic products, and lack the land it would take to grow them. Farmers in developing countries often see their crops rot in the fields long before they can be eaten or rushed across rutted dirt roads to markets many hours away. To those people, the Western cult of organic food is nothing more than a glorious fetish of the rich world—one with the power to kill them.

IT'S HARD to find anything positive to say about Thomas Malthus. After all, his dour view of the world has consistently been proven wrong. In 1798, he argued that the earth's population was rising exponentially and the food supply necessary to feed it was not. He famously promised "famine . . . the last, the most dreadful resource of nature." It took another 125 years for the world's population to double, but only fifty more for it to double again. Somehow, though, the food supply remained adequate. Mass starvation has often seemed inevitable, yet it has almost always been averted. Why is that? How could Malthus, not to mention the many apostles of doom who followed, have been so wrong? The answer is simple: science and technology have repeatedly saved humanity.

Over the past two hundred years there has been a progression of technological leaps that Malthus could never have imagined. The astonishing advances in human welfare, and in our ability to address poverty, have largely been the result of the discovery of effective antibiotics and vaccines for more than a dozen deadly diseases. But steam power, steel plows, and agricultural success were no less essential. We simply get far more out of every crop than seemed possible when Malthus was alive. The result has been one of the modern world's greatest achievements: good food for billions of people at prices that got lower every year.

By 1940, however, that system was beginning to fail in many countries: Mexico, China, India, and Russia all seemed on the verge of famine. Even parts of Europe were threatened. Once again, experts braced themselves for the worst. "In the 1970s and 1980s hundreds of millions of people will starve to death in spite of any crash programs embarked upon now," Paul Ehrlich wrote in *The Population Bomb*, which was published in 1968. "I have yet to meet anyone familiar with the situation who thinks India will be self-sufficient in food by 1971." He also insisted that "India couldn't possibly feed two hundred million more people by 1980." Once again, technology—and human imagination—interceded (as it has for hundreds of thousands of years, at least since some distant ancestor turned a stone into an ax). Between 1960 and 1985—during which time Erlich published his book and the Club of Rome issued *The Limits to Growth*, its modern echo of Malthus's grim assessment—food production in most of the world's poorest countries more than doubled.

The revolution began as a single experiment by one man, Norman Borlaug, an American plant scientist, who while working in

Mexico had spent years crossing the local wheat with Japanese dwarf varieties to produce plants that could respond better to irrigation and benefit more consistently from fertilizer. That approach was quickly applied to corn, beans, and rice, and the results could soon be seen planted across hundreds of millions of acres throughout Latin America and Asia. The impact is still hard to believe: despite a 300 percent increase in the population of the earth since the end of the Second World War, by far the fastest such growth spurt in human history, available calories per capita have risen by nearly 25 percent. India not only survived the 1960s (with help from the United States) but has since seen its population double, its wheat production triple, and its economy grow ninefold. India has also become one of the world's biggest rice exporters. The lives of billions of people have been transformed.

All technological advances have costs. Many are painful and most are unanticipated. The Green Revolution was no exception. With little thought devoted to land management and driven by an almost limitless reliance on water, the environmental impact has been staggering. For decades, India and China have been digging wells and damming rivers from one end of Asia to the other. The dams have displaced millions. Wells have liberated a generation of farmers from their dependence on rain, but clean water doesn't flow forever. As the population grows, particularly in the world's two most populous countries, the freshwater dwindles, and that leaves people with just one choice: dig. Drill too deep, though, and saltwater and arsenic can begin to seep into the ground, and when that happens nothing will grow on that land again.

For the first time since 1960, we are in a race to see whether the planet can provide enough food to feed its inhabitants. There are really only two ways to increase the amount of food a country can produce. Either you coax greater yield out of land already devoted to farming, or you find extra space to grow more. Historically, agriculture has alternated between the two strategies; for the past century, however, there has been a lot of both. Today, crops are grown on nearly 40 percent of the earth's land, and it takes 70 percent of our water to do it. Farming is, by its nature, an assault on the earth. Tilling, plowing, reaping, and sowing are not environmentally benign activities and they never were. Moreover, it has been estimated that pests, viruses, and fungi reduce agricultural productivity throughout the world by more than a third. You can't turn a crop into edible food without killing pests. And you can't kill them without poison—whether man-made or natural.

There is only so much war you can wage on your environment, however, and we have just about reached our limit. Physical expansion is no longer a meaningful option because we have run out of arable land. Three-fourths of the farmland in sub-Saharan Africa, where a third of the population suffers from chronic hunger, has become nutritionally useless, and more than 40 percent of the African continent suffers from desertification. "Globally, we're losing soil at a rate twenty times faster than it is formed," writes David R. Montgomery, a professor of geomorphology at the University of Washington and author of the 2007 book *Dirt: The Erosion of Civilizations*. Montgomery estimates that farming is responsible for eroding as much as 1 percent of the earth's topsoil every year. If that doesn't change we could literally run out of soil within a century.

By 2050, if not sooner, the earth will have half again as many people as it does today, more than nine billion. Long before that, though, possibly within the next twenty years, world food demand will have doubled. The Green Revolution largely bypassed Africa, and people in many countries there are actually getting poorer; but something surprising has happened throughout much of the rest of the developing world. Success itself has placed unbearable new burdens on the food supply. Agrarian societies have traditionally consumed little meat. But in China and other East Asian nations where income has been growing rapidly, that is no longer true. In India, 65 percent of the population still works on farms. Nonetheless, the country now has more than 280 million urban residents, and the shift to city life, which began more than a hundred years ago as rural residents fled famine and drought, continues.

Every day one hundred thousand Indians join the middle class; the trend in China is similar. As people get wealthier, and as they move away from the farm, their eating habits change. The biggest of those changes is that they start to eat meat. The UN's Food and Agriculture Organization (FAO) expects that global meat production will double by 2050 (which is more than twice the rate of human population growth). The supply of meat has already tripled since 1980: farm animals take up the vast majority of agricultural land and eat one-third of the world's grain. In the rich nations we consume three times the meat and four times the milk per capita of people in poorer countries. But that is changing rapidly, and as it does we will have to find ways to grow more grain to feed those animals—and to do it all on less land, and with less available water, than we have today. It is *this* demographic reality, more than population growth alone, that most seriously threatens the global food system.

Climate change, environmental degradation, water scarcity, and agricultural productivity are all intertwined. It will not be possible to solve any of those problems unless we solve them all. Climate change is likely not only to bring warmer temperatures but also to alter patterns of rainfall, placing even more stress on agriculture. Livestock already consume 80 percent of the world's soybeans and more than half the corn. Cattle require staggering amounts of fresh, potable water. It takes thirteen hundred gallons of water to produce a single hamburger; a steak requires double that amount.

Water scarcity may be the most visible problem caused by our addiction to meat, but it is not the only one: to make a pound of beef requires nearly a gallon of fuel. To put that into perspective, producing one kilogram of the grass-fed beef so revered by organic devotees and high-end restaurants causes the same amount of greenhouse gas emissions as driving a small car 70.4 miles. Even for beef raised less luxuriously (fed by grain on industrial farms) the figure is nearly forty-five miles. Eating meat is ecologically ruinous: according to a 2008 study by researchers at Carnegie Mellon University, if we all skipped meat and dairy just one day each week it would do more to lower our collective carbon footprint than if the entire population of the United States ate locally produced food every day of the year.

Malthus may have badly underestimated human ingenuity, but he did get one formula right: combine intense population pressure with high levels of poverty, reduce the opportunity for technological advances, and the guaranteed result will be famine and death. In 2005, an average hectare of land could feed four and a half people; by 2050 that same plot will need to support at least six people (and possibly closer to eight). The only way that will happen is by producing more food per hectare—more crop, as agronomists like to say, per drop. That is not the direction in which the world has been moving. Grain production began to decline in the 1990s for the first time since World War II. Africa, the continent that needs the most help, is the place that is faltering most profoundly. Total production on farms there, according to the World Resources Institute, is nearly 20 percent less than it was in 1970. Without another agricultural revolution, that trend will surely accelerate.

IF WE GENUINELY care about sharing our fate, and making food more readily available to *everyone*, there is only one question worth asking: how can we foment that next revolution? Certainly we need a better way to grow crops, one that sustains the earth but also makes the most efficient possible use of it. Breeding is the art of choosing beneficial traits and cultivating them over time. Farmers have done that for thousands of years by crossing plants that were sexually compatible and then selecting among the offspring for what seemed like desirable characteristics—large seeds, for example, or sturdy roots. That had always been a laborious and time-consuming process: mixing vast numbers of genes—sometimes whole genomes—almost entirely at random meant transferring many genes agronomists didn't want in order to get the ones they were looking for.

These extra genes often had negative effects, and it could take years of testing new strains to remove them. It was a crude system, akin to panning for tiny amounts of gold in a rushing river filled with stones, but given enough time it usually worked. By conserving seeds and careful mating, farmers learned how to make better plants, as well as entirely new varieties. All the plants we eat (corn, wheat, peanuts, rice) and many that we don't (orchids, roses, Christmas trees) have been genetically modified through breeding in an effort to make them last longer, look better, taste sweeter, or grow more vigorously in arid soil. So have most varieties of grapefruit, watermelon, lettuce, and hundreds of other fruits, vegetables, and grains that are for sale in any supermarket.

Evolution, which works on a different time scale and has no interest in easing the life of any particular species, does essentially the same thing: selects for desired traits. Humans have no choice but to try and hasten the process. Modern agriculture—and modern medicine—really didn't begin until 1953, when James Watson and Francis Crick discovered the structure of the DNA molecule, which carries the information that cells need to build proteins and to live. Genetics and molecular biology are simply tools to help scientists choose with greater precision which genes to mix (and how to mix them).

Advocates of organic farming, almost always speaking from—if not for—the world's richest countries, say the "natural" approach to breeding plants could solve food shortages and address issues of environmental sustainability at the same time. More importantly, they argue that genetic engineering has promised more than it can, or at least has, delivered (which is true, in part because opposition and bureaucratic meddling have made it true). The most vocal criticism of genetically engineered crops, and the easiest to dismiss, is based on willful ignorance, the driving force of denial-ism. The best-known representative of this group is Prince Charles, who summed his argument up nicely many years ago: "I happen to believe that this kind of genetic modification takes mankind into realms that belong to God, and to God alone." Putting aside the fact that not all farmers believe in God, the prince's assessment betrays his complete ignorance of the continuum of evolution and the unmistakable connection between "conventional" plant breed-ing and genetic engineering.

All the foods we eat have been modified, if not by genetic en-gineering then by plant breeders or by nature itself. After all, corn, in its present form, wouldn't exist if humans had not cultivated the crop. The plant doesn't grow in the wild and would never survive if we suddenly stopped eating it. Does God object to corn? The prince skipped over another, equally essential truth: genetic mutation occurs naturally in all living things. Genes are constantly jumping around and swapping positions without any laboratory assistance; in fact, evolution depends on it.

There are more legitimate reasons to worry about genetically engineered foods. The speed with which this technology has spread across the globe transformed agriculture before many people ever realized it. "So confident are the technicians of the safety of their products that each one is seen as no more than an arbitrary mix of independent lengths of DNA," the popular British geneticist Steve Jones has written. "Their view takes no account of the notion of species as interacting groups of genes, the properties of one . . . depending upon the others with which it is placed." Virus-resistant crops, for example, contain viral genes in all their cells. But viruses can introduce genetic material to their host cells, which means that these crops could, in theory, be able to create new diseases rather than defend against them.

The most vivid example of this kind of unintended consequence occurred in 1995, when scientists working at the seed company Pioneer Hi-Bred placed genes from a Brazil nut into a soybean, to help increase levels of two amino acids, methionine and cysteine, in order to make beans used as animal feed more nutritious. Technically, the experiment was a success, but the newly engineered bean also demonstrated how changing just a few molecules of DNA might affect the entire food chain. Many people are allergic to Brazil nuts, and they pay particular attention to labels. Yet labels cannot list every amino acid used to cultivate every crop that is then eaten by every animal, and which might ultimately find its way into a product. If somebody were unwittingly to eat a cake made with soy that contained the Brazil nut protein, the results could be deadly. (In this case, the Brazil nut soybean was never eaten. Pioneer took blood from nine people in a laboratory, and stopped the experiments when the serum tested positive. Still, with such research occurring in countries that lack strong regulatory systems, similar mistakes could have frightening consequences.)

There is an even darker and more abiding fear: that genetically engineered pollen will escape into the wild, altering plant ecosystems forever. That is both more likely and less dangerous than it seems. Pollen doesn't simply plop onto any plant, have sex, and create new seeds; it would first have to blow across a field and land on a compatible mate. If not, there would be no new seeds and little environmental danger. Genetically engineered crops have been planted on more than one billion acres, yet there have

been no examples of domesticated crops damaged by genetic pro-
miscuity. That doesn't mean it couldn't happen—but it's not surprising that it hasn't. Most major crops have few relatives close
enough to mate with, and wild species don't mix easily with those
that are domesticated.

Biotechnology is not without risks for people or the environment, nor is its potential unlimited. Nonetheless, that potential
can never be expanded or explored as long as irrational fear and
zealous denial prevent nearly every meaningful attempt to introduce genetically engineered crops in places like Africa. Agricultural investment and research there has withered even as the
population continues to climb. European and American critics
frequently state that the risks of genetically engineered crops outweigh their benefits. They have unrealistic expectations—as denialists so often do. If people in Geneva or Berkeley want to pretend
that genetically engineered products pose a danger that scientists
have been unable to discover, they should go right ahead. The risk
and reward equation looks entirely different in sub-Saharan Africa, however, where starvation is common and arable land almost
impossible to find.

No continent needs agricultural improvement more desperately
than Africa; yet there is no place where fear and denialism are more
pronounced. (Until recently South Africa was the only country
that permitted the harvesting of genetically engineered crops for
commercial uses; not long ago Kenya became the second.) Why
the resistance? Some leaders simply reject Western products on
principle, particularly those, like drugs and engineered crops, that
are hyped as vehicles of salvation. Commerce, too, plays a role, and
so does history. "The governments and citizens of Europe continue

to exercise considerable postcolonial influence in Africa through a range of mechanisms," Robert Paarlberg wrote in his 2008 book *Starved for Science: How Biotechnology Is Being Kept Out of Africa.* Paarlberg, who has long studied the impact of science and technology on farmers in the developing world, noted that European countries provide a great deal of technical assistance, financial aid, and nongovernmental advocacy to Africa. But nothing comes without strings attached, and African governments learned quickly that nobody in European countries had any intention of purchasing exports grown with modified seeds. "Through each of these channels today Europe is telling governments in Africa that it would be best to stay away from agricultural GMOs and African governments have responded accordingly," Paarlberg wrote.

Total reliance on organic farming would force African countries to devote twice as much land per crop as we do in the United States. It would also put the profligate West in the position of telling the world's poorest nations—as well as its fastest-growing economies—that they don't deserve to reap benefits that we have for so long taken for granted (and abused). That is the central message agricultural denialists have for Africa, and not just for Africa. It may be possible to convince China and India that burning less coal in their factories will not only ease carbon emissions but also lower their considerable health care costs. Lecturing people who have just purchased their first car or apartment about how cheeseburgers are going to kill them or destroy the planet is a different task. Do as I say, not as I do doesn't work with American teenagers; why should it work in Bangalore or Beijing? It should be no surprise that McDonald's franchises are growing faster in India and in China than anywhere else in the world.

Growth and poverty have come together—often in the same countries—to threaten the future for us all. Nearly a billion people go to bed hungry every night. Lack of food is not the only reason, and some argue that it is not even the principal reason. Politics, war, greed, tribal hatred, and bad government also contribute significantly to the problem. Now, so does something else: the growing demand for agricultural feedstocks to use as biofuel. In 2008, that demand pushed food prices ever higher, despite the recession, and the number of starving people rose to 14 percent of the world's population, according to the Food and Agriculture Organization. Three out of every four of those people live in rural areas and depend on agriculture to stay alive. As the world's financial crisis deepens, the bleak international economy can only add to the suffering. (Even lower prices rarely help struggling farmers during a severe recession; they are simply left with fewer incentives to plant a new season's crops. At the same time, poor people are finding it nearly impossible to obtain loans to buy seed and fertilizer.)

To cope, Africans will need better governments. The quality of farming doesn't really matter to countries engaged in eternal civil war or riven by corruption. The continent will also have to acquire new technology and the skill to employ it aggressively. "We are going to need a lot of inventiveness about how we use water and how we grow crops," Nina Fedoroff said. Fedoroff, a molecular biologist who has worked on plant genetics for many years, is the science adviser to Secretary of State Hillary Clinton. She believes people have become so hobbled by their fear of genetically engineered food that it threatens not only progress but peace. "People clearly are afraid and that is very hard to watch," she said.

become plentiful and affordable to all. And that is a feat that we are not conscious of. But it has changed the world."

Agricultural work has always been—and remains today—humanity's principal occupation. In the United States we long ago ceased to know, or care, where most of our food comes from. We have managed to liberate ourselves so completely from that dangerous and demanding life that the Census Bureau no longer bothers to count in a separate category the number of people who live on farms. Never before has the responsibility to feed humanity been in the hands of so few, and never before have so many been oblivious of that fact.

Americans rarely think about wheat when they eat bread (let alone know how to bake it), and most have never seen a live pig or chicken, unless they have visited a zoo. Poultry and pork are not animals; they are standard parts of a meal—born and raised in a baggie. There has been an understandable reaction to this mechanized approach to our food. The organic movement—fueled by rural nostalgia and pastoral dreams—is one that shuns mass production, stresses tradition, and seeks to return to less complicated times when the land was tilled by simple farmers, not regulated by computers and planted under the care of the Global Positioning System. Call it the Old McDonald fantasy.

The desire may be genuine, but it is based on a dangerous fallacy: that the old days were better. We can think that as long as we don't have to live it—because it is true nowhere. The old days were treacherous and painful. Nasty, brutish, and short was the rule, not the exception. Life expectancy two centuries ago in Europe and America was little more than half of what it is today. Science changed all that, helping to feed the poorest people on earth; it brought farmers throughout Asia and Latin America a new kind of prosperity.

Today the world, and particularly its poorest inhabitants, needs more science, not less. Much of the technology Africa requires has been available—to us—for decades. Without passable roads, products never make it to market. Modern irrigation systems are almost wholly absent from Africa, but they would permit farmers to grow larger crops with less water, as such systems have done nearly everywhere else in the world. More than anything, Africa needs soil that has an adequate supply of nitrogen. Without nitrogen fertilizers we would lose a third of our crops. Organic evangelists argue that the best way to get more nitrogen into Africa is to use more manure. Clearly, these are people who have never been to Tanzania or Tamil Nadu. Africans and Indians have plenty of manure. In fact, human and animal waste are often the only farming resources available to villagers. They have no other choices. In effect, that means organics have been imposed on them. Cuba is an interesting illustration; the country has often been portrayed as an organic utopia because it had no genetically engineered crops or synthetic fertilizers. Nobody could afford them. In 2009, however, the government announced that it was about to plant its first crop of engineered corn. With access to technology the ideological barriers vanished. The consequences of any other approach would be horrific. Feeding the world with organic food would require vast new tracts of farmland. Without ripping out the rain forests, there just isn't enough of it left.

At the TED conference, Fresco was displaying pictures of African farmers, people for whom securing daily meals is no more certain than it was for our Paleolithic ancestors. "If we want small-scale farming we will relegate these farmers and their families to poverty," she said. "What they need are implements to increase their production. Fertilize their soil. Something to protect their crops. Small-scale farming is a luxury." She peered knowingly into the crowd. "A luxury for those of us who can afford it."

I WAS ONCE invited to dinner by a friend who ate nothing but organic food. We picked up vegetables on the way to her house: broccoli, squash, and peppers. Then we bought swordfish. When we arrived at her house, my friend walked straight into the backyard, fired up her Weber grill, sliced the vegetables, and proceeded to cook them. "I just do it this way," she said, "so they don't lose their vitamins."

Vitamins are good for you; but cancer isn't. Charred food contains carcinogens; so does charcoal and the grease that often drips from a grill into the fire. The food we ate that night was far more likely to cause harm than any conventional product cooked another way. The genuine risks never occurred to her. Like many people, though, she buys organic food because it makes her feel safer. But there is no such thing as safer. There is only safer than something else. Skiing and driving cars are thousands of times more dangerous than walking or cycling. Yet we never refuse to enter a motor vehicle because it "may" cause death. In most parts of America, tap water is not at all dangerous, a fact that is well publicized. That hasn't put a dent in the bottled water industry.

When people decide that science can't solve their problems, they reject its principles. Denying the truth becomes a habit. First we say, oh, pesticide is causing illness, so I'll eat only organic food. Or perhaps chemicals are the problem. The solution for that is simple: use only natural medicine. Lord Melchett, of the British Soil Association, put it this way: "It will be consumers, *not scientists*, who decide whether pesticide residues are safe to consume." So much for the value of facts or the idea of objective standards. Why bother assessing the safety of foods or employing scientists at all? Nearly every day there seem to be new and contradictory directives about what to eat and how to eat it. For some people the most coherent response is to say, in one way or another, "Civilization causes cancer," so they begin to turn their backs on civilization.

Many producers of organic food have seized on that fear and uncertainty, advertising their goods as natural and healthy alternatives to this intangible and remote system of corporate farms. Never mind that nearly all the organic crops in the United States are grown or sold by the same food conglomerates that grow and sell conventional produce. Giant corporations like Heinz, Cargill, Kellogg, and Kraft have gobbled up organic food companies throughout the nation. Why wouldn't they? If customers are willing to pay twice as much for foods cultivated without synthetic pesticides or that lack genetically modified ingredients, Kellogg and General Mills will be only too happy to sell it to them.

Thoughtful proponents of precaution argue that at least with organic crops we know what is likely to happen. Genetically engineered products are so new that we can't be sure. "No one person or group knows or understands enough about the complexity of living things or their intimate interactions or what affects them to declare that biotechnology and genetic engineering are risk-free," Denise Caruso wrote in her book *Intervention:*

Confronting the Real Risks of Genetic Engineering and Life on a Biotech Planet. "In fact, the only thing we all share—scientists, citizens, regulators—is the profound uncertainty of this moment in history."

She is completely right. But has there ever been a meaningful new technology that carried no risk, or that couldn't be used for bad as well as for good? Francis Bacon recognized the answer to that question four hundred years ago. "It would be an unsound fancy and self-contradictory to expect that things which have never yet been done can be done except by means which have never yet been tried," he wrote in *The New Organon.* I wonder what he would have made of the "precautionary principle," which holds that potential risks, no matter how remote, must be given more weight than any possible benefit, no matter how great. Without accepting some risk we would never have had vaccines, X-rays, airplanes, or antibiotics. Caution is simply a different kind of risk, one that is even more likely to kill people.

In Europe, the caution industry suffocates innovation. In Rome, where I lived for several years in the 1990s, they refer to organic food as "biological" and look upon genetically engineered crops as unadulterated poison. No sane person would swallow it willingly. And the Italian government helps ensure that they won't. To sell their seeds to farmers, companies must present a certificate stating that their products have not been genetically engineered. At harvest time, farmers are required to do that, too, as are food processors and supermarket chains. America may seem more tolerant, but actually the food system is just less heavily regulated. If Louise Fresco had held up two ears of corn at TED—one grown organically and the other engineered with a toxin to resist worms and fungus—I am certain the voting would have been no different than it was with the bread.

"Just the mention of genetic engineering, a process that has been used for thirty years and so far has not harmed a single person or animal, can cause alarm," Pamela C. Ronald has pointed out in *Tomorrow's Table: Organic Farming, Genetics, and the Future of Food*, which she wrote with her husband, Raoul Adamchak. The two make an unusual couple: she is professor of plant pathology and chair of the Plant Genomics Program at the University of California at Davis. He is an organic farmer. Perhaps not surprisingly, they believe agriculture can—and must—accommodate both approaches. This makes them the agronomic equivalent of James Carville and Mary Matalin—a couple who represent camps defined by their mutual hostility.

Tomorrow's Table is a brilliant, though perhaps futile, attempt to reconcile the warring sides. "The apocalyptic quality of the anti-GE advocacy seems wildly disproportionate to the potential risk, particularly in the context of benefits," Ronald wrote in the book. "Unlike fluoride or some types of synthetic or organic pesticides such as rotenone"—an odorless organic chemical found in the roots and stems of many plants—"which are unquestionably lethal to animals at high concentrations, GE traits are composed of the same chemical building blocks (DNA and proteins) that we eat every day. Indeed, these are the same components that Buddha ate 2,500 years ago, and they are what we will be eating 2,500 years from now."

The National Academy of Sciences and the United Kingdom's Genetically Modified Science Review Panel, among many other scientific organizations, have concluded repeatedly that the process of adding genes to our food by genetic engineering is just as safe as conventional plant breeding. Each group, in turn, has concluded that there is no danger associated with replacing the combination of genes that has always occurred through breeding (or nature) with a process that allows scientists to insert snippets of DNA into the walls of cells with a gene gun.

If scientific consensus mattered, there would be little debate about whether to use our most promising technology to help feed billions of people who have no reasonable alternative. Nor would there be much question that genetically engineered crops, which require fewer and less-toxic chemicals, are at least as good for the environment as organic crops that guzzle more water per acre and require up to seven times as much herbicide. The amount of pesticides used on corn, soybeans, and cotton in the United States has declined by more than 2.5 million pounds since genetically engineered crops were introduced in 1996, according to one study funded by the Department of Agriculture. In addition, the herbicide glyphosate—more commonly known as Roundup—is less than one-third as toxic to humans than the herbicides it replaces. It is also far less likely to persist in the environment.

This type of manipulation has long been accepted in medicine, largely because the risks seem minor and the benefits easy to understand. Insulin produced since 1982, for example, has been made from a synthetic gene that is a replica of one found in humans. Nobody seems to have problems with cancer or heart drugs based on biotechnology either. Yet altering the molecular genetics of the food supply remains a boundary that many people are unwilling to cross.

The opposition is so uniform and reflexive that when in 2004 the FAO issued a carefully prepared and comprehensive report that dared to suggest that "agricultural biotechnology has real potential as a new tool in the war on hunger," nongovernmental organizations throughout the world rose as one to object. Six hundred and fifty groups banded together, signing an open letter in which they said that the "FAO has broken its commitment to civil society and peasant organizations." The letter went on to complain that groups representing the interests of farmers had not been consulted, that the FAO was siding with the biotechnology industry, and, consequently, that the report "raises serious questions about the independence and intellectual integrity of an important United Nations agency."

This type of response was hardly an aberration. The attack on Iowa governor Tom Vilsack, Barack Obama's agriculture secretary, began the day Obama announced his nomination. Vilsack's crime, according to the Organic Consumers Association, was extreme. (The OCA describes itself as the "only organization in the US focused exclusively on promoting the views and interests of the nation's estimated 50 million organic and socially responsible consumers.") Vilsack believes in biotechnology at least as fully as the leaders of the OCA believe in organic food, and that automatically makes him suspect. Typically, organizations like the OCA denounce any official who supports genetic engineering, no matter what the reason. Most of Iowa's farmers grow genetically engineered foods, and they wouldn't have it any other way. Vilsack's central transgression was that, as governor, he considered that a good idea.

Attacks on progress have become routine. Look at these comments from a group whose members refer to themselves as "independent scientists" at the Third Joint International GMO Opposition Day, April 8, 2006: "The current generation of genetically modified crops unnecessarily risks the health of the population and the environment. Present knowledge is not sufficient to safely and predictably modify the plant genome, and the risks of serious side-effects far outweigh the benefits. We urge you to stop feeding the products of this infant science to our population and ban the release of these crops into the environment where they can never be recalled."

Not one fact in any of those sentences is true. While 70 percent of all processed food in the United States contains at least one ingredient from genetically modified corn, canola, or soybeans, beyond using the word "billions," it is not possible to guess with any accuracy how many doses of such food Americans have actually consumed in the past thirty years. But it *is* possible to count the number of people who have become ill as a direct result of eating that food: zero. Not one. Nearly two thousand Americans died after taking aspirin in 2008 (out of twenty-nine billion pills swallowed), and another three hundred drowned in their bathtubs. Aspirin sales haven't suffered, and people are still taking baths.

"NATURAL" DOES NOT mean good, or safe, or healthy, or wholesome. It never did. In fact, legally, it means nothing at all. Mercury, lead, and asbestos are natural, and so are viruses, *E. coli*, and salmonella. A salmonella outbreak in 2009 killed nine people, sickened hundreds, and triggered the largest food recall in the history of the United States, sending a chill through every parent who has ever made a peanut butter and jelly sandwich. Other than mosquitoes, the two substances responsible for more deaths on this planet than any other are water and "natural" food. Wine and beer were invented as ways to purify water and make it safer to drink; the fermentation process destroys many of the most dangerous pathogens. If the Chinese had not understood the importance of boiling water for tea, they would have been sipping cups full of deadly fungi and other dangerous pathogens for the last five thousand years.

Organic food almost always explicitly excludes the use of genetic engineering or synthetic chemicals. "Natural" chemicals and pesticides are far more common, and no safer, however, than chemicals made in any laboratory. As James E. McWilliams, author of *American Pests: The Losing War on Insects from Colonial Times to DDT,* has written, "One issue frequently overlooked in the rush to embrace organic agriculture is the prevalence of excess arsenic, lead, cadmium, nickel, mercury, copper, and zinc in organic soil. Soil ecologists and environmentalists—and, to some extent, the concerned public—have known for more than a century that the synthetic pesticides of conventional farming leave heavy metals in the ground. But the fact that you'll find the same toxins in organic soil has been something of a dirty little secret."

While the risks of genetically modified foods are constantly cited, the dangers of nature are rarely mentioned. As the Berkeley biochemist Bruce N. Ames has demonstrated, a single cup of coffee contains more natural chemicals than most people will consume in a month of eating three daily meals. That doesn't mean coffee is dangerous. It just means nature makes lots of chemicals, and they are no less toxic than those made by man. When invoking studies of toxicity, organophiles often tell only one side of the story. (Which, of course, is a hallmark of denialism.) Any chemical, whether it comes from the root of a tree or the shelves of your medicine cabinet, can cause serious harm. It depends how much you take. That is why one of the fundamental tenets of medicine holds that "the dose makes the poison."

For decades, plant breeders and farmers have routinely blasted crops with radiation. The practice, mutagenesis, is not organic, but has been widely—and quietly—accepted throughout the world as a way to hasten the breeding of plants. Even those who wouldn't eat irradiated food rarely object publicly as they do with genetically engineered products. Mutagenesis produces new hybrids at remarkable speeds, but it also causes rapid mutations in their genetic structure. Seeds are typically collected, germinated, and surveyed for new traits.

In 2008, a team of plant geneticists based in Portugal published a report that compared the effects brought on by this type of radiation with those caused by genetic engineering. They examined the protein structure of four strains of rice, focusing on the nutrients, toxins, and allergens contained in thousands of their genes. Without exception, the changes induced by mutagenesis were more significant than any brought about using the tools of molecular biology. Again, that doesn't mean mutagenesis is dangerous. It's not. Surely, though, radiation—a process that effects the entire plant—ought to frighten people more than the manipulation of a single gene. Yet nobody has ever refused to let a ship dock at an African port because it was filled with irradiated wheat. (In 2002, 2.4 million Zambians faced starvation. Nevertheless the government rejected as "poison" tons of genetically engineered grain offered by the World Food Program.)

Does organic food carry a lower environmental footprint than food grown with the use of synthetic pesticides? The answer to that is complicated but it certainly isn't yes. Locally grown food has environmental benefits that are easy to understand. Agricultural researchers at Iowa State University have reported that the food miles—the distance a product travels from farm to plate—attached to items that one buys in a grocery store are twenty-seven times higher than those for goods bought from local sources. American produce, every cauliflower or radish, travels an average of nearly fifteen hundred miles before it ends up at our dinner table. That doesn't make for fresh, tasty food and it certainly doesn't ease carbon emissions.

People assume that food grown locally is organic (and that organic food is grown locally). Either may be true, but often neither is the case. It's terrific news that Michelle Obama has decided to grow vegetables at the White House; her family will eat better, not because the food is organic, but because it will be fresh. Go to a nearby farmers' market and buy a tomato or apple that was grown by conventional means. It will taste good if it was recently picked. Then buy an apple from the organic section of your local supermarket. It will have been grown according to standards established by the U.S. Department of Agriculture: no synthetic pesticides, no genetic manipulation.

That doesn't mean it was picked when it was ripe. If those organic apples aren't local, they ripened while they were stored—

usually after having been sprayed with ethylene gas to turn them red from green (ethylene is one of the many chemicals permitted under the USDA's contradictory and mystifying organic guidelines). The British Soil Association rules permit the use of ethylene too, as a trigger for what it refers to as "degreening" bananas. The association says that it's acceptable to use ethylene in the ripening process for organic bananas being imported into Europe, in part because "without a controlled release of ethylene bananas could potentially ripen in storage." In other words, they would begin to undergo the organic process known as rotting.

Food grown organically is assumed to be better for human health than food grown in conventional ways. Recent studies don't support that supposition, though. In 2008, for example, researchers funded by the Danish government's International Center for Research in Organic Food Systems set out to look at the effect of three different approaches to cultivating nutrients in carrots, kale, peas, potatoes, and apples; they also investigated whether there were differences in the retention of nutrients from organically grown produce. The crops were grown in similar soil, on adjacent fields, and at the same time so that they experienced the same weather conditions. The organic food was grown on organic soil, but it was all harvested and treated in the same manner. The produce was fed to rats over a two-year period. Researchers, led by Susanne Bügel, an associate professor in the department of human nutrition at the University of Copenhagen, reported in the *Journal of the Science of Food and Agriculture* that the research "does not support the belief that organically grown foodstuffs generally contain more major and trace elements." Indeed, she and her team found no differences in the nutrients present in the crops after harvest, and no evidence that the rats retained different levels of nutrients depending on how the food was grown.

If organic food isn't clearly better for the environment or our health, if it doesn't necessarily carry lower carbon costs or cost less money, will people stop buying it? Probably not. Look at what's happening with milk. Pasteurization has made dairy food safe enough to serve as one of the foundations of the American diet. Raw milk is legal in nearly half the states, however, and it is easy to buy in the others. There are raw milk clubs, furtive Web sites, and clandestine milk-drinking clubs all over America today. I spoke to a "dealer" at the Union Square farmers' market in New York one morning. He didn't actually have the "stuff" with him, but he was willing to arrange a meet.

This would be funny if it wasn't deadly. As the University of Iowa epidemiologist Tara Smith has reported on her blog, Aetiology, several deaths and more than a thousand illnesses have been linked to raw milk consumption between 1998 and 2005 in the United States—a tenfold increase from the previous decade. And the business is booming. "Raw milk is like a magic food for children," said Sally Fallon, president of the Weston A. Price Foundation, a group that supports the consumption of whole, natural foods. Its advocates claim that raw milk relieves allergies, asthma, autism, and digestive disorders. No data exists to support any of those assertions.

There is plenty of data associated with the consumption of raw milk, however. In 1938, for example, milk caused 25 percent of all outbreaks of food- and water-related sickness in the United States. Universal pasteurization brought that figure to 1 percent by 1993, according to the Center for Science in the Public Interest, a nutrition advocacy group in Washington. "It's stunning," Marion Nestle said to me, "to think that so many people have decided to reject one of the most successful public health achievements we have had in the past century. It really makes you wonder what people want and who they trust."

CANOLA—AN ACRONYM for Canadian oil, low acid—has been around for less than fifty years. A derivative of the ancient rapeseed plant, canola has some attractive properties, including a lower level of saturated fat than most oils, and a rich supply of omega-3 fatty acids. In 2009, the German chemical company BASF introduced a strain that is resistant to a particular class of herbicides called imidazolinones. Douse the crop and almost magically the weeds die while the canola remains unharmed. Engineering crops to do that has been the biggest priority for biotechnology firms.

When freed to use a single effective spray that kills weeds without harming their crops, farmers need less herbicide. That saves money and helps the environment. (In China, during 1997, the first year cotton resistant to the bollworm was introduced, nearly half a billion dollars was saved on pesticides. More importantly, cotton farmers there were able to eliminate 150 million pounds of insecticide in a single year. As Pamela Ronald has pointed out, that is nearly the same amount of insecticide as is used in California every year.)

Monsanto introduced the herbicide Roundup in 1996. Roundup Ready seeds, which were engineered to resist that herbicide, have dominated every market in which they are sold. Yet their very ability to tolerate chemicals has provoked controversy. When used excessively (and improperly), Roundup can linger in fields long after it has done its job. The same is true for organic herbicides, but there is at least one difference: genetically engineered crops are scrutinized in a way that no other food has ever been. That won't happen to the new canola from BASF, though, because scientists bred the mutation it needs to resist herbicides without relying on the techniques of biotechnology. In other words, they did "naturally" what genetic engineering does in a lab. And to opponents of genetically engineered food that makes all the difference.

Nature hasn't noticed. Despite its pedigree, the BASF canola seems to pose more of a threat to the environment than any crop from a test tube. "Some crops cannot be planted in the year after" the herbicides are sprayed on the new canola strain, according to University of Melbourne plant geneticist Richard Roush. In test sites, he found residues lingering in the soil at levels far greater than any caused by Roundup or similar herbicides. "From an agronomic standpoint," he told the *New Scientist*, "it has all the issues of genetically modified canola seed, but it is arguably worse."

A plant bred in a laboratory is no more or less "real" than a baby born through in vitro fertilization. The traits matter, not the process. A crop doesn't know if it emerged after a week of molecular research or three thousand years of evolution. The new strain of canola is not yet available commercially; when it is, European farmers will be able to plant it anywhere they like. If it were the product of biotechnology, however, European regulations would prevent farmers from using it at all. Nobody should assume that a food is safe because it has been genetically engineered. But should we honestly accept assertions that organic food is more socially progressive than food made with chemical herbicides? Or that raw milk possesses healing powers?

Change is hard to accept, and change for no apparent reason is especially upsetting. Purple tomatoes and fluorescent fish seem freakishly unnatural. (Snapdragon genes placed in tomatoes cause their skin to turn dark purple. Glofish that have fluorescent genes come in "three stunningly beautiful colors," according to the company that makes them: "Starfire Red®, Electric Green®, and Sunburst Orange®.") There is a difference between psychedelic fish and essential foods, however. After years of opposition, the French government declared in 2009 that a single strain of corn modified to resist the European corn borer, which has been harvested on millions of acres around the world without causing harm, was safe enough to plant and eat. Not everyone in the French government concurred, however. The environment minister, Jean-Louis Borloo, announced that he had no intention of lifting the ban because it would pose too great a threat.

Fear of genetically engineered foods has warped some of the very principles that environmentalists hold most sacred: that resources should be conserved, and the earth farmed wisely. Bt, for example, is an insecticide derived from the spores and toxic crystals of the bacterium *Bacillus thuringiensis* and even organic farmers spray it on their plants. Place the gene inside the plant, however, and it becomes unacceptable (taking us, as Prince Charles would have it, into "God's realm"). One recent study in northern China, though, demonstrated that genetically engineered cotton, altered to express the insecticide Bt, not only reduced pest populations among those crops, but also among others nearby that had not been modified with Bt. It can't destroy every pest, but no insecticide comes closer.

For many people the scariest thing about genetically modified crops has nothing to do with science. It's about their seeds. In enormous swaths of the world, seeds are heritage. People are often paid with them, and they conserve them more carefully than almost any other asset. Often there *are* no other assets. When a company like Monsanto comes along selling one type of corn seed, which can only be controlled with one particular insecticide that Monsanto also happens to make—well, who can compete with that? When they sell seeds that cannot reproduce, people become even more alarmed, fearing that they might be forced every year to buy their crop again (at prices over which they have no control).

"This is an argument I have never understood," Robert Shapiro said. Shapiro, who is retired as Monsanto's chairman, became the Johnny Appleseed of genetically modified foods. He was also the embodiment of his company's failed efforts to market those products to Europe (and beyond). To environmentalists, Shapiro has long been seen as Satan, which is ironic, because until they started threatening his life Shapiro was a card-carrying member of Greenpeace. "We weren't eliminating any choice that was already available," he said, explaining his approach to transgenic crops. "If people didn't want to buy the stuff, they could keep doing exactly what they're doing, no one was taking anything away from anybody. It was just, if you want to use the new technology, then you have to use it on a set of conditions. And if you don't think that's a good deal, keep on doing what you're doing. It didn't make you any *worse* off." Shapiro saw seeds in much the same way that Bill Gates thought about software: as a form of intellectual property.

Food isn't software, however, and farmers throughout the developing world became genuinely terrified of losing their livelihoods, particularly when so much of the world's engineered seed is controlled by a few giant corporations. Nevertheless: what happened with software has increasingly become true in agriculture as well. If you don't want to use Windows or Word or Excel (or can't afford them), there are excellent alternatives. Some are cheaper and many are free. That's the power of the open-source approach to intellectual goods. Even if Monsanto had wanted to control the world's grain, the company could never have succeeded: farmers save and share seeds, and in countries like Bangladesh and India national seed-breeding programs have been instituted to make sure people can get seed they can afford. There are open-source grains and cheap public seed banks in many developing countries. Over half the rice planted in China now is hybrid, and farmers buy it every year—usually from local seed companies.

Genetic engineering is what many environmentalists refer to as a "corporate technology" because it has mostly been used by industrial agricultural conglomerates to provide benefits to farmers and residents of rich countries. That has been true. People in the Loire Valley or Cambridge, Massachusetts, don't need a tomato that resists frost or ripens only after a week. Products like that are not going to save the world, maybe not even help the world. As the British economist Michael Lipton put it to me years ago, "I always say that electricity is a fantastic invention, but if the first two products had been the electric chair and the cattle prod, I doubt that most consumers would have seen the point."

It has taken years—even decades—to develop genetically engineered organisms that serve the poor. A store full of beneficial foods, such as cancer-fighting carrots and rot-resistant fruits, does not exist. Neither does an AIDS vaccine; should we give up on that? It is not an idle comparison, because the people who have the most to gain from medical and agricultural biotechnology are Africans. Neither Monsanto nor Syngenta has invested heavily in improving the yields of cassava, yams, rice, or bananas. But honestly, why should they? What incentive could they possibly have? In 1986, pharmaceutical companies abdicated much of the American vaccine market, because lawsuits made it impossible for them to profit. They don't spend much money trying to cure visceral leishmaniasis either. It's a parasite we don't get in Manhattan, and the millions in the Third World who do suffer from it can't pay for the treatment.

The market doesn't solve every problem. That is one reason why the Bill and Melinda Gates Foundation has spent billions of dollars to vaccinate children who could never otherwise see a doctor. As it happens, the foundation (and others) has now embarked on a similar program that focuses solely on cassava, which is the primary source of calories for nearly a billion people—250 million of whom live in sub-Saharan Africa. The crop has many deficiencies: it is almost entirely made of carbohydrates so it cannot provide balanced nutrition to people who subsist on it; once harvested the plant must be processed quickly or it will generate poisonous cyanide within days. The roots deteriorate rapidly, which limits the food's shelf life; and Gemini virus, a common plant disease, destroys up to half of every harvest.

An international team led by Richard Sayre, a professor of plant cellular and molecular biology at the Ohio State University, has been working feverishly to overcome every one of those problems. Sayre calls it the most ambitious plant genetic engineering project ever attempted. The team has already succeeded with individual traits, though in separate plants. The researchers introduced genes that can facilitate mineral transport and help the roots draw more iron and zinc from the soil. They have also reported a thirtyfold increase in the levels of vitamin A, which is critical for vision. Soon they will attempt to create a single plant that expresses all the traits. Sayre said he hopes to have the fortified cassava tested in Africa by 2010.

That's one example. There are many others. Scientists are working on plants that resist drought and salt and others that can shrug off the most common but deadly viruses. After years of scientific struggles and bureaucratic interference, golden rice, which carries genes that make it possible to produce beta-carotene, which is then broken down into vitamin A, is about to enter the food chain. The World Bank estimates that in India alone, golden rice could save as much as the equivalent of 1.5 million years of life every year. Critics of the product have tried to block it for years (so far with great success), arguing that the rice isn't needed and won't work. According to a report released in 2008 by the International Federation of Organic Agricultural Movements, an adult would have to eat nine kilograms of cooked golden rice a day to absorb the minimum daily requirement of vitamin A. That was true a decade ago, but science has moved forward. The new generation of golden rice is more efficient and nine kilograms have been reduced to 150 grams, which is not too much to digest and well within the economic reach of even the poorest people.

The earth isn't utopia and never will be—but insisting that we can feed nine billion people with organic food is nothing more than utopian extremism, and the most distressing and pernicious kind of denialism. An organic universe sounds delightful, but it would consign millions of people in Africa and throughout much of Asia to malnutrition and death. That is a risk everyone should be able to understand.

"Even if the worst thing anyone imagines about genetically modified organisms were true, they would be worth it," said William C. Clark, a professor of international science, public policy, and human development at Harvard University. Clark has spent much of his career trying to figure out the most environmentally benign way to feed the world. "If you look at what people are dying of in Africa and what these plants could do to produce food, we would have to be absolutely out of our mind not to use them. You could triple the risks. Make them the worst risks imaginable. Even then, it wouldn't be a contest."

4

The Era of Echinacea

Not long ago, for reasons I still don't understand, I began to
feel unfocused and lethargic. Work was no more stressful
than it had ever been, and neither was the rest of my life. The bulk
of my savings had been sucked into the vortex of the newly recognized black hole called the economy. But whose had not? I try
to eat properly, exercise regularly, sleep peacefully, and generally
adhere to the standard conventions of fitness. It didn't seem to be
working. My doctor found nothing wrong and my blood tests
were fine. Still, I felt strange, as if I were lacking in energy—or in
something. So I did what millions of Americans do every day. I
sought salvation in vitamins.

First, though, I had to figure out what variety of salvation to
seek. There are many thousands of pills, potions, powders, gels,

elixirs, and other packaged promises of improved vitality for sale within just a few blocks of my home. I walked to the closest store, a place called The Health Nuts, and told the proprietor I was feeling sluggish. He nodded gravely and took me straight to the amino acid section. To counteract my deficit of energy, he recommended a supplement of glutamine, which is one of the few amino acids that passes the blood-brain barrier. When people are under stress—physical or psychological—they begin to draw down on their stores of glutamine. "This stuff repairs brain cells and it's good for depression, too," he told me. A leaflet attached to the bottle described the amino acid as a magical aid for mental acuity. ("It is helpful with focus, concentration, memory, intellectual performance, alertness, attentiveness, improving mood, and eliminating brain fog & cloudiness"). I dropped it into my basket.

The store also had a garlic section—not actual garlic, but various pills with names like Kyolic and Garlicin, GarliMax and Garlique, all of which claimed to possess the healing properties of garlic, which for centuries has been thought to help ward off the common cold, clear up respiratory infections, and soothe sore throats. Garlic, its advocates claim, is also effective in treating heart problems, lowering cholesterol, and keeping arteries free of blood clots. I grabbed a bottle and moved on to the main supplement section, where multivitamins in every conceivable size, shape, dosage, strength, and formulation were lined up in rows. (There were vitamins for vegans, and for people allergic to gluten, for those who don't need iron and those who do; and there were specific pills for every age group, from the fetus right through to the "well-derly.") Antioxidants were next to them, all seemingly fueled by the "natural" power of prickly pear, goji, and açaí, the intensely popular Brazilian berry that supposedly offers benefits such as rejuvenation, skin toning, and weight loss, not to mention prevention of various illnesses like heart disease. There was also something called "BlueGranate," a combination of blueberries and pomegranates, both of which "possess wondrous health properties," as the bottle put it. "A synergistic blend of powerful and potent phytonutrient antioxidants." Into the basket it went.

Almost everything advertised itself as an antioxidant. Oxidation is a natural result of metabolic processes that can cause harmful chain reactions and significant cellular damage. Those broken cells in turn release unstable molecules called free radicals, which are thought to be the cause of many chronic diseases. Set loose, free radicals can turn into scavengers, ransacking essential proteins and DNA by grabbing their electrons for spare parts. Antioxidants prevent those reactions, but standing there, it was impossible to know how, or if, they worked. The collection of pills was so enormous, the choice so vast, and the information so humbling that while I may not have been depressed when I arrived at The Health Nuts, spending half an hour there did the trick.

I went home and consulted the Internet, which was even more intimidating: there are millions of pages devoted to vitamins and dietary supplements. You could spend your life combing through them and then another life trying them all out. Fortunately, my eye was drawn immediately to the Vitamin Advisor, a free recommendation service created by Dr. Andrew Weil, the ubiquitous healer, whose domed head and bearded countenance are so profoundly soothing that with a mere glance at his picture I felt my blood pressure begin to drop.

Dr. Weil is America's most famous and influential practitioner of complementary medicine—he prefers to call it integrative—which seeks to combine the best elements of conventional treatment with the increasingly popular armamentarium of alternatives, everything from supplements to colonic irrigation, spiritual healing, and homeopathy.

The public's hunger for novel remedies (and alternatives to expensive drugs) has transformed the integrative approach into one of the more potent commercial and social forces in American society. Nearly every major medical school and hospital in the country now has a department of integrative or complementary medicine. (A few years ago the Harvard Medical School even tussled with its affiliate, the Dana Farber Cancer Institute, over which would win the right to house such a program. Harvard prevailed.) While the movement has grown immensely since he opened his Center for Integrative Medicine in Arizona in 1994, Weil remains at its heart. Educated at Harvard University, both as an undergraduate and at its medical school, Weil embraces herbal therapies, New Age mysticism, and "spontaneous healing," which is the title of one of his books. But he also understands science and at times even seems to approve of it.

Weil offers sound advice in his many books—calling refined foods, excess starches, corn sweeteners, and trans fats dangerous, for example, and noting that exercise and a proper diet are far more beneficial even than the vitamins and supplements he recommends. His influence is immense, and in a country embarking on an urgent debate about how to make its health care system more affordable, rational, and responsive, that influence has never been felt more powerfully. Weil is in great demand as a public speaker, testifies before Congress, and has twice appeared on the cover of *Time* magazine. For advocating the many health benefits of mushrooms, Weil is a hero to mycologists the world over. (He is one of the rare Americans to have had a mushroom named after him, *Psilocybe weilii*.)

Andrew Weil seemed like just the man to lead me out of the forest of nutritional darkness into which I had inadvertently wandered. His Vitamin Advisor Web site assured me that, after answering a few brief questions, I would receive "a personalized comprehensive list of supplements based on my lifestyle, diet, medications, and health concerns"—all at no cost, without obligation, and prepared specially to meet my "unique nutritional needs." In addition, if I so chose, I could order the "premium quality, evidence-based" supplements in the proper doses that would "exactly match the recommendations from Dr. Weil." The supplements would be "custom packed in a convenient dispenser box and shipped directly to [me] each month." The only thing Dr. Weil doesn't do for you is swallow the pills.

I filled out the form, answering questions about my health and providing a brief medical history of my family. Two minutes after I pressed "submit," Dr. Weil responded, recommending a large number of dietary supplements to address my "specific health concerns." In all, the Vitamin Advisor recommended a daily roster of twelve pills, including an antioxidant and multivitamin, each of which is "recommended automatically for everyone as the basic foundation for insurance against nutritional gaps in the diet." Since, as he points out on the Web site, finding the proper doses can be a "challenge," Dr. Weil offered to "take out the guesswork" by calculating the size of every pack, which, over the previous ninety seconds, had been customized just for me. That would take care of one challenge; another would be coming up with the $1,836 a year (plus shipping and tax) my new plan would cost. Still, what is worth more than our health? If that was how much it would cost to improve mine, then that was how much I was willing to spend.

Also on my list: milk thistle, "for those who drink regularly or have frequent chemical exposure," neither of which applies to me; ashwagandha, an herb used in ayurvedic medicine to help the body deal with stress and used traditionally as an energy enhancer; cordyceps, a Chinese fungus that for centuries has been "well known" to increase aerobic capacity and alleviate fatigue; and eleuthero, also known as Siberian ginseng, often employed to treat "lethargy, fatigue and low stamina." In addition, there was 1000 milligrams of vitamin C (high doses of which, the information said, *may* provide additional protection against the oxidative stress of air pollution and acute or chronic illness); saw palmetto complex, mixed with stinging nettle root to "support prostate health"; an omega-3 pill (which *may* help reduce the symptoms of a variety of disorders); Saint-John's-wort (to support healthy mood); folic acid (which in addition to offering pregnant women proven protection against neural tube defects, *may* have a role to play in heart health, and *may* also help protect against cancers of the lung and colon, and even *may* slow the memory decline associated with aging); and finally another ayurvedic herb, triphala (a mixture of three fruits that help tone the muscles in the digestive tract).

Dr. Weil, who argues that we need to reject the prevailing impersonal approach to medicine, reached out from cyberspace to recommend each of these pills wholeheartedly and specifically, just for me. Before sending off a check, however, I collected some of the information on nutrition and dietary health offered by the National Institutes of Health, the Harvard School of Public Health, and the Memorial Sloan-Kettering Cancer Center. It turned out that my pills fell essentially into three categories: some, like cordyceps and triphala, seemed to do no harm but have never been shown in any major, placebo-controlled study to do any particular good; others, like Saint-John's-wort, may possibly do some good in some cases for some people, but can also easily interfere with and negate the effects of a large number of pre-scribed medicines, particularly the protease inhibitors taken by many people with AIDS. Most of the pills, however, including the multivitamin and antioxidant, seemed just plain dangerous.

Despite Dr. Weil's electronic assurances that his selections were "evidence-based," not one of those twelve supplements could be seen to hold anything more than theoretical value for me. At best. One study, completed in 2008, of the omega-3 fatty acids so ben-eficial when eaten in fish, found that in pill form they had no discernible impact on levels of cholesterol or any other blood lip-ids. The study was not large enough to be definitive; other trials are needed (and already under way). But it would be hard to argue with Jeffrey L. Saver, vice chairman of the American Heart Asso-ciation's Stroke Council, professor of neurology at UCLA, and the director of its department of Stroke and Vascular Neurology, who called the findings "disappointing."

Others agreed. "You know, most of that stuff just comes right out at the other end," former surgeon general C. Everett Koop told me. The ninety-four-year-old Koop is congenitally incapable of ignoring facts or pretending they shouldn't matter. "Selling snake oil has always been one of America's greatest con games. But the more we know about our bodies, the more people seem to buy these pills. That part I never did understand; you would have hoped it would be the other way around. But every day it becomes clearer: we need to eat properly and get exercise. And every day more people seem to ignore the truth."

New data keeps streaming in, and almost all of it confirms that assessment. In 2009, researchers from the Women's Health Initiative, working at dozens of major medical centers under the direction of the National Heart, Lung, and Blood Institute, concluded a fifteen-year study that focused on strategies for preventing heart disease, various cancers, and bone fractures in postmenopausal women. After following 161,808 women for eight years, the team found no evidence of any benefit from multivitamin use in any of ten conditions they examined. There were no differences in the rate of breast or colon cancer, heart attack, stroke, or blood clots. Most important, perhaps, vitamins did nothing to lower the death rate.

Another recent study, this time involving eleven thousand people, produced similar results. In 2008, yet another major trial, of men, had shown that the risk for developing advanced prostate cancer, and of dying from it, was in some cases *actually twice as high* for people who took a daily multivitamin as it was for those who never took them at all. There are hundreds of studies to demonstrate that people who exercise regularly reduce their risk of coronary artery disease by about 40 percent, as well as their risk of stroke, hypertension, and diabetes, also by significant amounts. Studies of vitamin supplements, however, have never produced any similar outcome.

Antioxidants, often described in the press as possessing wondrous powers, and recommended to every American by Dr. Weil, among others, are taken each day by millions. That should stop as soon as possible. While a diet rich in antioxidants has been associated with lower rates of chronic disease, those associations have never been reflected in trials in which people took antioxidants in supplement form. In 2007, for example, the *Journal of the American Medical Association* published the results of the most exhaustive review yet of research on such supplements. After examining sixty-eight trials that had been conducted during the previous seventeen years, researchers found that the 180,000 participants received no benefits whatsoever. In fact, vitamin A and vitamin E, each immensely popular, actually increased the likelihood of death by 5 percent. Vitamin C and selenium had no significant effect on mortality. (Vitamin C has long been controversial. Linus Pauling, the twentieth century's greatest chemist, was convinced it would cure cancer. He was wrong. In fact, too much vitamin C actually seems to help cancer cells withstand some kinds of treatment.)

"The harmful effects of antioxidant supplements are not confined to vitamin A," said the review's coauthor, Christian Gluud, a Danish specialist in gastroenterology and internal medicine and head of the trial unit at the Centre for Clinical Intervention Research at Copenhagen University Hospital. "Our analyses also demonstrate rather convincingly that beta-carotene and vitamin E lead to increased mortality compared to placebo." More than a quarter of all Americans over the age of fifty-five take vitamin E as a dietary supplement, yet among healthy people in the United States it would be hard to cite a single reported case of vitamin E deficiency.

It gets worse: folic acid supplements, while of unquestioned value for pregnant women, have been shown to increase the likelihood that men would develop prostate cancer. "Unfortunately, the more you look at the science the more clearly it tells you to walk away," Kelly Brownell said. Brownell, director of the Rudd Center for Food Policy and Obesity at Yale University, has for years studied the impact of nutrition on human health. "Vitamins in food are essential. And that's the way to get them. In food." With a couple of exceptions like folic acid for pregnant women, and in some cases vitamin D, for the vast majority of Americans dietary supplements are a complete waste of money. Often, in fact, they are worse.

That brings us back to Dr. Weil, who understands the arguments against using vitamin supplements in a country where, with rare exceptions, people have no vitamin deficiencies. He actually makes them pretty well in his book *Healthy Aging: A Lifelong Guide to Your Physical and Spiritual Well-Being*. "Not only is there insufficient evidence that taking [antioxidants] will do you any good, some experts think they might be harmful," he wrote. Excellent analysis, pithy and true. In fact, the evidence of harm keeps growing. In May 2009, researchers from Germany and the United States reported in the *Proceedings of the National Academy of Sciences* that antioxidants like vitamins C and E actually reduce the benefits of exercise. "If you promote health, you shouldn't take large amounts of antioxidants," said Michael Ristow, a nutritionist at the University of Jena, who led the international team of scientists. "Antioxidants in general . . . inhibit otherwise positive effects of exercise, dieting and other interventions." Despite news like that, Dr. Weil still thinks you need to take his ("I continue to take a daily antioxidant formula and recommend it to others as well").

Weil doesn't buy into the idea that clinical evidence is more valuable than intuition. Like most practitioners of alternative medicine, he regards the scientific preoccupation with controlled studies, verifiable proof, and comparative analysis as petty and one-dimensional. The idea that accruing data is simply *one way* to think about science has become a governing tenet of the alternative belief system. The case against mainstream medicine is simple, repeated often, and, like most exaggerations, at least partially true: scientists are little more than data collectors in lab coats, people wholly lacking in human qualities. Doctors focus on disease and tissues and parts of the anatomy that seem to have failed, yet they act as if they were repairing air conditioners or replacing carburetors rather than attending to the complex needs of an individual human being. And pharmaceutical companies? They serve no interest but their own. Complementary and alternative medicine, on the other hand, is holistic. It cares. In the world of CAM, evidence matters no more than compassion or belief. Weil spells it all out in *Healthy Aging*:

> To many, faith is simply unfounded belief, belief in the absence of evidence, and that is anathema to the scientific mind. There is a great movement toward "evidence-based medicine" today, an attempt to weed out ideas and practices not supported by the kind of evidence that doctors like best: results of randomized controlled trials. *This way of thinking discounts the evidence of experience.* I maintain that it is possible to look at the world scientifically and also to be aware of nonmaterial

reality, and I consider it important for both doctors and patients to know how to assess spiritual health. (Italics added.)

Evidence of experience? He is referring to personal anecdotes, and allowing anecdotes to compete with, and often supplant, verifiable facts is evidence of its own kind—of the denialism at the core of nearly every alternative approach to medicine. After all, if people like Weil relied on the objective rules of science, or if their methods were known to work, there would be nothing alternative about them. If an approach to healing has a positive physical effect (other than as a placebo), then it leaves the alternative world of sentiment and enters the world of science and fact. The only attribute that alternatives share is that they do not meet the scientific standards of mainstream medicine.

Data is not warm or kind. It is also, however, not cold or cruel. Assessing data and gathering facts are the only useful tools we have to judge whether a treatment succeeds or fails. Weil understands that, and yet he mixes perfectly sensible advice with lunacy. ("I would look elsewhere than conventional medicine for help if I contracted a severe viral disease like hepatitis or polio, or a metabolic disease like diabetes," not to mention "treatment for cancer, except for a few varieties.")

That makes him a uniquely dangerous proponent of magical thinking. It is much easier to dismiss a complete kook—there are thousands to choose from—than a respected physician who, interspersed with disquisitions about life forces and energy fields, occasionally has something useful to say. Still, when Weil writes about a "great movement toward 'evidence-based medicine'" as if that were regrettable or new, one is tempted to wonder what he is smoking. Except that we don't need to wonder. He tells us.

Weil believes in what he calls "stoned thinking" and in intuition as a source of knowledge. This he juxtaposes with "straight" or "ordinary" thinking. You know, the type weighed down by silly rules and conventional thought. Like every alternative healer, Weil believes in the supremacy of faith and compassion. I certainly wouldn't argue against faith (if only because for many people it provides the single form of alternative medicine that seems clearly to work, a placebo effect). And here is my definition of compassion: the desire to alleviate suffering. Nothing in the course of human history meets that definition so fully as the achievements of evidence-based, scientifically verifiable medicine.

The world of CAM is powered by theories that have almost never been tested successfully, and its proponents frequently cite that fact as proof of their unique value, as if they represent a movement that cannot be confined (or defined) by trivialities. It would be terrific if Weil were correct when he says that evidence-based medicine is now in vogue; given its astounding record of success, it certainly ought to be. There is at least one compelling reason that the scientific method has come to shape our notion of progress and of modern life. It works.

But pendulums swing in more than one direction. As Steven Novella, director of general neurology at the Yale University School of Medicine, has written, the biggest victory won by proponents of complementary and alternative medicine was the name itself. "Fifty years ago what passes today as CAM was snake oil, fraud, folk medicine, and quackery," he wrote on Neurologica, his blog, which is devoted heavily to critical thinking. "The pro-moters of dubious health claims were charlatans, quacks, and con artists. Somehow they managed to pull off the greatest con of all—a culture change in which fraud became a legitimate al-ternative to scientific medicine, the line between science and pseudoscience was deliberately blurred, regulations designed to protect the public from quackery were weakened or eliminated, and it became politically incorrect to defend scientific standards in medicine."

The integrity of our medical system is certainly subject to doubts and debate. Doctors can be smug and condescending, and they often focus on treating diseases rather than preventing them. But in the alternate universe of CAM treatment nobody has to prove what is safe, what works and what doesn't. And that's dangerous because Americans are desperate for doctors who can treat their overall health, not just specific illnesses. Perhaps that is why this particular universe, populated by millions of peo-ple, has been fueled by one of American history's most unlikely coalitions—the marriage of the extreme right with the heirs of the countercultural left.

The political right has never wavered in its support for dietary supplements, and Orrin Hatch, the Utah Republican, has long been the industry's most powerful supporter. Hatch doesn't share a lot of political space with Tom Harkin, the populist liberal from Iowa. They don't agree on abortion rights, gun control, or many other issues. But when it comes to the right of every American to swallow any pill he or she can find in a health food store, the two are welded by a bond of steel. "For many people, this whole thing is about much more than taking their vitamins," Loren D. Israelsen, an architect of the 1994 legislation that deregulated the supplement industry, said. "This is really a belief system, almost a religion. Americans believe they have the right to address their health problems in the way that seems most useful to them. Often, that means supplements. When the public senses that the government is trying to limit its access to this kind of thing, it always reacts with remarkable anger—people are even willing to shoulder a rifle over it. They are ready to *believe anything* if it brings them a little hope."

That kind of fervent belief, rather than facts, feeds disciplines like ayurvedic medicine, which argues for the presence of demonic possession in our daily life, and Reiki, the Japanese practice of laying on the hands, which is based on the notion that an unseen, life-giving source of energy flows through each of our bodies. Then there is iridology (whose practitioners believe they can divine a person's health status by studying the patterns and colors of his iris), Healing—or Therapeutic—Touch, qi gong, magnet therapy. None of it works. Acupuncture, while effective in reducing arthritic pain and the impact of nausea, has never been demonstrated to help people quit smoking or lose weight—two of its most popular applications.

Homeopathy, perhaps the best-known alternative therapy, is also the most clearly absurd, based as it is on the notion that "like cures like." In other words, it presumes that a disease can be treated by ingesting infinitesimally small dilutions of the substance that caused the disease in the first place. No matter what the level of dilution, homeopaths claim, the original remedy leaves some kind of imprint on the water molecules. Thus, however diluted the solution becomes, it is still imbued with the properties of the remedy. No homeopathic treatment has ever been shown to work in a large, randomized, placebo-controlled clinical trial, but nothing seems to diminish its popularity. With logic that is both ridiculous and completely sensible, the federal government has taken a distant approach to regulating homeopathy *precisely because* it contains no substance that can possibly cause harm (or good). "Homeopathic products contain little or no active ingredients," Edward Miracco, a consumer safety officer with the FDA's Center for Drug Evaluation and Research explained. As a result, "from a toxicity, poison-control standpoint" there was no need to worry about the chemical composition of the active ingredient or its strength.

On those rare occasions when data relating to alternative medicine does become available, it is almost invariably frightening: in 2004, for example, a large group of researchers reported in the *Journal of the American Medical Association* that more than 20 percent of the ayurvedic medicines the group purchased on the Internet contained detectable and dangerous levels of lead, mercury, and arsenic. Soon afterward, the FDA warned consumers to exercise caution when purchasing ayurvedic products.

"Who is telling people that all this stuff is good?" asked Anthony S. Fauci, the director of the National Institute of Allergy and Infectious Diseases. "Their peers. The scientists say either we don't know or it doesn't work. And their response is the same thing that they're doing with the vaccine issue. They want to feel better, they want to live forever, they want not to age. So people start going around with all of these herbs that are not proven to work and that the scientists are skeptical about. Which all the more, I think, makes them want to go after it. There's an element of, 'I'll show you, you son of a bitch. The people who I hang out with think this really works.' And then you come out with a paper, like the one in the *New England Journal of Medicine*, that shows there is absolutely no benefit from echinacea. Bingo! They don't care. They don't care a bit."

The Obama administration's laudable desire to bring medical costs under control and to make the health care system more accessible has presented leaders of the CAM community with a unique opportunity—and they have seized it. In February 2009, Weil and other famous supporters of natural healing, including Mehmet Oz, a cardiac surgeon and founder of the Complementary Medicine Program at New York Presbyterian Hospital (and, most famously, Oprah Winfrey's health guru); and Dean Ornish, of the Preventive Medicine Research Institute in Sausalito, California, testified before the Senate Health, Education, Labor, and Pensions Committee and correctly pointed to disease prevention as the key to crafting the new health care legislation President Obama has committed himself to so completely. In their testimony, however, each did battle with a series of straw men, arguing that the American health care system costs too much (it does), is too reliant on technology (it is), and is geared to complicated and costly treatment rather than to prevention, which is cheaper and almost always more effective (also true). A month before the summit, the group published a piece in the *Wall Street Journal* in which they wrote, "Our 'health-care system' is primarily a disease-care system. Last year, $2.1 trillion was spent in the U.S. on medical care, or 16.5% of the gross national product. Of these trillions, 95 cents of every dollar was spent to treat disease *after* it had already occurred. At least 75% of these costs were spent on treating chronic diseases, such as heart disease and diabetes, that are preventable or even reversible."

None of that is really in dispute, but those fundamental flaws cannot possibly be overcome by a system that replaces facts with wishes. Prescribing diet and exercise to fight disease is *not* an alternative approach to medicine, as anyone who has visited a physician in the past five years would surely know. What America needs, and what the Obama administration has, for the first time, set out to do, is get a better sense of which treatments work and which don't. That would, for instance, require placing caloric information on all fast-food menus and explaining what it means. And it will require clear economic judgments about whether many current procedures are in fact worth the cost. All that will require data, not voodoo.

Nearly a decade ago, a Stanford University professor named Wallace I. Sampson warned that institutional support for alternative medicine endangers society. "Modern medicine's integrity is being eroded by New Age mysticism, cult-like schemes, ideologies, and classical quackery," he wrote in an influential essay called "The Alternate Universe," arguing that they were all misrepresented as "alternative" medicine. "Using obscure language and misleading claims, their advocates promote changes that would propel medicine back five centuries or more. They would supplant objectivity and reason with myths, feelings, hunches, and sophistry." At the time, Sampson's claims seemed a bit over the top to me. We were in a golden era of medicine; life expectancy grew nearly every year, and so did our knowledge of how to treat many chronic diseases. It never occurred to me that science-based medicine might be considered an obstacle to a healthy life, rather than the best chance of having one.

Those bottles of folic acid and BlueGranate from The Health Nuts were sitting on my desk. They looked so promising and appeared to offer so much: support for a healthy cardiovascular system, as well as better memory and brain function; they would promote urinary tract, eye, and skin health, boost body detoxification functions, and reduce cellular damage associated with the aging process. There was, however, a tiny asterisk next to each claim. "These statements have not been evaluated by the Food and Drug Administration," each one said. "These products are not intended to treat, diagnose, cure or prevent any disease." I looked at the Garlicyn, the amino acids, and the vitamin C. Same warnings. Same nearly invisible print.

The fine print on labels like that is rarely read. In the world of dietary supplements, facts have always been optional. Pharmaceuticals are strictly regulated; for supplements there is almost no oversight at all. A pill may be sold as something that contains 1000 mg of vitamin C. But how can you be sure without meaningful standards? When facts are not required anything goes, and Andrew Weil, for one, wouldn't have it any other way. "I believe in magic and mystery," he wrote in *Healthy Aging*. "I am also committed to scientific method and knowledge based on evidence. How can this be? I have told you that I operate from a *both-and mentality, not an either-or* one."

Sorry, but that's not possible. Either you believe evidence that can be tested, verified, and repeated will lead to a better understanding of reality or you don't. There is nothing in between but the abyss. The FDA knows that, and so does the supplement industry. And so does Andrew Weil. If a product whose label promotes it as contributing to brain function, cardiovascular health, or one that can reduce cellular damage associated with aging, or improve digestion, or support a healthy immune system is "not intended to cure, treat, diagnose or even prevent" any health problem, what on earth, one has to wonder, is it supposed to do?

ALMOST 40 PERCENT of American adults made use of some form of alternative medical therapy in 2007, according to the most recent *National Health Statistics Reports.* They spent $23.7 billion on dietary supplements alone. It has become one of the America's biggest growth industries. (And one that almost uniquely profits during times of economic distress. People are far more likely to turn to herbs and other supplements when they can't afford genuine medical care—and when they have no access to any other health system.) There were approximately 4,000 supplements on the market in 1994, when the industry was deregulated by Congress. Today the exact number is almost impossible to gauge, but most experts say there are at least 75,000 labels and 30,000 products. Those numbers don't include foods with added dietary ingredients like fortified cereals and energy drinks, which seem to fill half the supermarket shelves in the country.

The attraction isn't hard to understand. For all that medicine has accomplished, millions of people still suffer the considerable aches and pains of daily life. Arthritis and chronic pain plague America, and much of that agony is no more amenable to pharmaceutical relief today than it was thirty years ago. The drugs one needs to alleviate chronic pain—aspirin, for instance—can cause their own complications when taken in high enough doses over a long enough period. The pharmaceutical industry is a monolith that often acts as if there is, or soon will be, a pill for everything that ails you. Too much cholesterol? We can melt it away. Depressed? Try one of a dozen new prescriptions. Can't sleep? Blood pressure too high? Obese, sexually dysfunctional, or bald? No problem, the pharmaceutical industry is on the case.

Even our medical triumphs cause new kinds of problems. Reducing deaths from heart disease and cancer, for example, permits us to live longer. And *that* exposes us to a whole new set of conditions, most notably Alzheimer's disease, a debilitating, costly, and humiliating illness for which there is no cure and few treatments of any value.

So what could it hurt to try something new? It is an era of patient empowerment. People have access to more information than ever, their expectations have changed, and they demand greater control over their own health. Supplements and herbal alternatives to conventional drugs, with their "natural" connotations and cultivated image of self-reliance, fit in perfectly. They don't require machines or complex explanations. People can at least try to relate to an herb like echinacea, which has been around for centuries, no matter how useless it is, or a practice like qi gong, which means "cosmic breathing" and suggests that human life forces can be marshaled to flow through our body in a system of "meridians." Homeopathy is nothing more than fraud, as any number of scientists, studies, reports, and institutions have pointed out. Yet, in a complex world simplicity offers an escape from the many moving parts of the medical machine. As with organic food, if science seems allied with corporations and conglomerates—all distant and unfathomable—well, then, nature feels just right.

Under the banner of natural and alternative treatments Americans reflexively accept what they would never tolerate from a drug company (and never should). Vioxx made that clear. Without post-marketing surveys, the unacceptable risks of Vioxx would never have been known. Maybe none of the tens of thousand of herbal supplements for sale in the United States carries any similar risk. But how would we know, since that kind of monitoring has never been required of supplements? Doctors could not have continued to prescribe Vioxx after news of its dangers was made public. Yet compare the way Vioxx was removed from the market—amid the greatest possible publicity and under threat from billions of dollars' worth of lawsuits—to what happened in 2004 with ephedra, which was America's most popular dietary supplement.

Ephedra, derived from the Asian herb ma huang, has been used for thousands of years; the herb's active ingredient, ephedrine, boosts adrenaline, stresses the heart, raises blood pressure, and is associated with an increased risk of heart attack, stroke, anxiety, psychosis, and death. None of that was in question. But the FDA's decision to pull it from the market certainly was. Many of the Americans who were outraged that Vioxx had been approved in the first place were just as outraged when ephedra was banned. It took the FDA years of legal battle to get the supplement removed from the shelves of vitamin shops. Ephedrine-containing supplements have caused deaths in many countries, not just in America. But people still want it, and what people want, the Internet provides. "We're growing ma huang (ephedra), which has been used in Traditional Chinese Medicine for 5,000 years but is now banned in America thanks to the criminally-operated FDA," one foreign supplier wrote in 2009. And he will be more than happy to sell it to anyone foolish enough to send money. And why would anyone do that? Because a supplement is not a drug. Its value is taken on faith and no amount of evidence will ever convince true believers to turn away.

Belief outranks effectiveness. Vitamin worship demonstrates that fundamental tenet of denialism with depressing regularity. In 2003, a study that compared the efficacy of echinacea to a placebo in treating colds received considerable attention. Researchers followed more than four hundred children over a four-month period, and found not only that a placebo worked just as well, but that children treated with echinacea were significantly more likely to develop a rash than those who took nothing at all.

Subsequent studies have been even more damning. In 2005, researchers from the Virginia School of Medicine reported in the *New England Journal of Medicine* that echinacea had no clinical impact, whether taken as a prophylactic or after exposure to a virus. Nor did it lessen the duration or intensity of any symptom. In addition, the American College of Pediatricians has urged parents to avoid echinacea mixtures for children who are less than a year old. The response? According to the latest data released by the federal government in 2008, echinacea remains the most heavily used supplement in the childhood arsenal. (It is still wildly popular with adults too, but fish oil is now in greater demand.)

Almost no restrictions were placed on the sale of supplements, vitamins, or other home remedies until 1906, when, reacting to the revelations in Upton Sinclair's book *The Jungle*, Congress passed the Pure Food and Drug Act. The law permitted the Bureau of Chemistry, which preceded the Food and Drug Administration, to ensure that labels contained no false or misleading advertising. Since then, the pendulum has swung regularly between unregulated anarchy and restrictions that outrage many Americans. In 1922, the American Medical Association made an effort to limit the indiscriminate use of vitamins, describing their widespread promotion as "gigantic fraud." It helped for a while. By 1966, the FDA tried to require the manufacturers of all multivitamins to carry this notice: "Vitamins and minerals are supplied in abundant amounts in the foods we eat. . . . Except for persons with special medical needs, there is no scientific basis for recommending routine use of dietary supplements." The vitamin industry made certain that no such warning was ever issued.

The relationship between food, drugs, and supplements began to blur in the 1970s as connections between diet, food, and medicine became more fully understood. What began as a federal effort to improve nutrition and prevent confusion has ended up as a tacit endorsement of chaos and deceit. First, with a major report issued in 1977 by the Senate Select Committee on Nutrition and Human Needs, and then with studies by the National Academy of Sciences and other research groups, the government started telling Americans to alter their diets if they wanted to have long and healthy lives. That made sense, of course. Advice about ways to reduce the risk of heart disease, diabetes, many cancers, and other chronic illnesses became routine. There were food pyramids and instructions to eat less salt and fat and add fiber as well as whole grains; eat more fruits and vegetables and watch the calories. Still, it was against the law to suggest that there was a relationship between the ingredients in a commercial food and the treatment or prevention of a disease.

Then, in 1984, came the Original Sin. That year, the National Cancer Institute lent its unparalleled credibility to the Kellogg Company when together they launched a campaign in which All-Bran cereal was used to illustrate how a low-fat, high-fiber diet might reduce the risk for certain types of cancer. All-Bran was the first food permitted to carry a statement that was interpreted widely as "Eating this product will help prevent cancer." That led to the era of product labels, and completely changed the way Americans think about not only foods but dietary supplements and ultimately about their health. Food was no longer simply food; it was a way to get healthy. Some of those changes made sense: flour was fortified with folate; juice enriched with calcium; and in 2004, in the name of health, General Mills started making every one of its breakfast cereals from whole grains.

They were exceptions. It would require Dickens's narrative skills and Kafka's insight into bureaucratic absurdity to decipher the meaning of most products for sale in American health food stores today. In the world of alternative medicine, words have become unmoored from their meanings. As long as a company doesn't blatantly lie or claim to cure a specific disease such as cancer, diabetes, or AIDS, it can assert—without providing evidence of any kind—that a product is designed to support a healthy heart, or that it protects cells from damage or improves the function of a compromised immune system.

It's still against the law to claim a product cures a disease—unless it actually does. But there is no injunction against saying that a food or supplement can affect the structure or function of the body. Such claims can appear on any food, no matter how unhealthy. You cannot advertise a product as a supplement that "reduces" cholesterol, but you can certainly mention that it "maintains healthy cholesterol levels." It would be illegal to state that echinacea cures anything, since of course it has been shown to cure nothing. But it's perfectly acceptable to say that echinacea is "an excellent herb for infections of all kinds," although no such thing has been proven to be true.

Even claims that are true are often irrelevant. Vitamin A, for example, is essential for good vision—as supplements for sale in any health food store will tell you. Insufficient consumption of vitamin A causes hundreds of thousands of cases of blindness around the world each year, but not in the United States; here people don't have vision problems arising from a lack of vitamin A. Although statements advertising vitamin A for good vision may, like many others, be legally permissible, they are meaningless. And since too much vitamin A can cause birth defects and osteoporosis, for example, its potential to harm American consumers is far greater than the likelihood that it will do good.

Not long ago, I was given a free bottle of Lifewater at my gym. "It's the perfect energy drink," the woman handing it out said, "because it's an antioxidant and nutritious. And of course, it's water." Except that Lifewater is not really any of those things. Water has no calories. My "Agave Lemonade Vitamin Enhanced Beverage" with natural flavors contained 40 calories per eight-ounce serving. That's five calories per fluid ounce, a little less than half the calorie count of regular Pepsi, the signature product of the corporation that sells Lifewater. Even that number, of course, is misleading. Nobody drinks eight ounces; Lifewater comes in a twenty-ounce bottle, which brings the calorie count to 100.

Okay, 100 calories is not that big a deal. But the main ingredient in the drink is sugar: 32 grams in what many people assume to be vitamin-enriched *water*. And that is not the kind of water we need to be drinking in a country where one-third of adults are obese, 8 percent are diabetics, and both of those numbers are rising rapidly. Lifewater contains no meaningful amount of agave, lemonade, yerba maté, or taurine, all of which are listed invitingly on the bottle.

Lifewater is hardly the only beverage created, named, or designed to dupe people into buying the opposite of what they are looking for; it's probably not even the worst offender. In January 2009, the Center for Science in the Public Interest sued the Coca-Cola Company in federal district court, saying that the company's Glacéau division relied on deceptive advertising and unsubstantiated claims when promoting VitaminWater as a "Nutrient-Enhanced Water Beverage" and by employing the motto "vitamins + water = all you need," for a product that has almost as much sugar in it as a similarly sized can of Coke. "VitaminWater is Coke's attempt to dress up soda in a physician's white coat," the CSPI litigation director Steve Gardner said when he filed the lawsuit. "Underneath, it's still sugar water, albeit sugar water that costs about ten bucks a gallon."

IT WOULD BE sufficiently distressing to know that the federal government permits unproven therapies to flourish in the United

States. Through the National Center for Complementary and Alternative Medicine, however, it actually encourages them. Josephine Briggs is an internationally renowned nephrologist who has published scores of scientific articles in prestigious journals. She served for nearly a decade as head of the Division of Kidney, Urologic, and Hematologic Diseases at the National Institute of Diabetes and Digestive and Kidney Diseases before moving, in 2006, to the Howard Hughes Medical Institute, where she had been named senior scientific officer. Briggs is agreeable personally, scientifically accomplished, and widely considered a talented administrator. It would be hard to imagine a less controversial choice to lead a branch of the National Institutes of Health.

Her appointment as director of the National Center for Complementary and Alternative Medicine in 2007, however, was not exactly greeted with hosannas. Many of her scientific peers regard the center as a distraction and a waste of money—or worse. (Indeed, NCCAM has been opposed since the day it was created, in part because of the way the center came to be: NCCAM is the brainchild of Iowa senator Tom Harkin, who was inspired by his conviction that taking bee pollen cured his allergies—a belief he has stated publicly before Congress. There is no evidence that bee pollen cures allergies or lessens their symptoms. For some people, however, it can cause life-threatening allergic reactions.)

Traditional scientists expressed hesitation about Dr. Briggs's appointment, but those were doubts about the center itself, not about her ability to do her job. The CAM community, on the other hand, saw themselves saddled once again with a "conventional" scientist as its leader. "New NIH NCCAM Director Wanted: No Experience or Interest in Field Required" ran the headline of one influential health blog, The Integrator, when word of the Briggs appointment surfaced in 2007. (The previous director, Stephen Straus, a renowned clinical virologist, died of brain cancer. He had also been criticized for his lifelong commitment to science-based medicine.) After Briggs was appointed, The Integrator was even more direct: "Oops, They Did It Again," the publisher, John Weeks, wrote early in 2008. In an open letter to Briggs he continued: "Director Zerhouni appointed you, despite the fact that you too have no visible professional experience in the field that you were selected to lead. Of your 125 publications, none appear to touch on the kind of interventions which will be on your desk at your new job. . . . Perhaps you can explain why the NIH would choose a novice, for the second time—it's officially 2 for 2 on this count—to run one of its domains? The answers that come to mind, and which I have heard from colleagues in the last 24 hours, range from fear to ignorance to suppression."

Briggs says she was prepared for opposition from both sides. It started the moment she told colleagues at the Howard Hughes Medical Institute about her new position. "Some of the criticism from conventional medicine arises from the usual sort of turf wars, and some of it arises from areas where there is genuinely some quackery," she said. "However, I think that we can steer this portfolio to the more mainstream practices, and the things that arouse the alarm of people who have been my colleagues all my scientific life are mostly things that are not being used by very many people." In her laboratory, Briggs has studied the effect of antioxidants on kidney disease, but she has never used alternative medicine, personally or in her practice. "I myself am not a supplement user," she said. "Vitamin D and calcium is my read of the literature"—by which she meant that those two supplements have proven their efficacy.

That's a sensible, fact-based approach to the current state of knowledge about dietary supplements. It is not, however, the "read" of the people who care passionately about CAM, including Senator Harkin, who established the center's precursor, the Office of Unconventional Medicine, in 1992 with $2 million in discretionary congressional funds. The word "unconventional" didn't sit well with the healing community, however, so the name was soon changed to Office of Alternative Medicine. Its mandate was simple: investigate treatments that other scientists considered a waste of time and money.

The first director, Joseph Jacobs, resigned under pressure from Harkin after he objected to including some nominees for the center's governing council. One of Harkin's choices had even endorsed the use of laetrile—perhaps the definitive quack treatment for cancer. (The laetrile movement, founded in the early 1950s by Ernst T. Krebs Sr., was based on the idea that a chemical found naturally in the pits of apricots could fight tumors. It couldn't, yet many desperate people spent the last days of their lives believing it would.) By 1998, the NIH director, Harold Varmus, a Nobel laureate who combined unimpeachable scientific credentials with a sophisticated understanding of Washington bureaucracy, wanted to place the Office of Alternative Medicine under more rigorous NIH scientific control. But Harkin wasn't about to let that happen; in a deft bit of administrative jujitsu, he managed to elevate the office into an independent research center within NIH.

That made the center even more suspect among scientists than it had been before. Briggs, like her predecessor, finds herself in an intellectually precarious position. In effect, she is like an attorney who argues that the guilt or innocence of her client is beside the point because everyone has the right to a defense. The defense most commonly offered in support of CAM is its stunning popularity—and popularity can easily be confused with reality. It happens every day on the Internet. Confusing popularity with authority is one of the hallmarks of denialism. People take comfort in becoming part of a crowd. The sense that complex issues can be resolved by a kind of majority vote, as if it was an election, helps explain the widespread support for CAM. It is also a reason the anti-vaccine movement has been so successful. Democracy rules. Millions of people take antioxidant supplements. That's unlikely to change. So the argument for NCCAM suggests that whatever we think of supplements, their popularity requires NIH to face reality and understand how they work.

Science doesn't operate by rules of consensus, however, and the NIH exists in order to discover scientific solutions to the health problems of Americans. Briggs understands that. "I think that there is a tension in this area, and I feel this as a practitioner—that while I'm comfortable in trying to reassure people, I'm not comfortable in fooling them. And you know, I think any physician is aware that some of the confidence you build in someone is part of helping them get better, and when is that being a confidence man? That tension is kind of inherent in healing practices. It's interesting; it's not easily solvable."

Briggs has become fascinated with the causes of the placebo effect—how it works on a biochemical level, and why. That the mind can affect the chemistry of the human body is not in doubt, and researchers have shown direct relationships between what a patient expects from a drug and its therapeutic results. In one experiment, Fabrizio Benedetti, professor of clinical and applied physiology at the University of Turin Medical School in Italy, demonstrated that a saline solution works just as well as conventional medicine to reduce tremors and muscle stiffness in people with Parkinson's disease. Benedetti is also a consultant for the Placebo Project at NIH and a member of the Mind-Brain-Behavior Initiative at Harvard University. In the Parkinson's study, he and his team found that neurons in patients' brains responded rapidly to saline. In another experiment, Benedetti has shown that for people who have no idea that a switch has been made, a shot of saline can provide as much pain relief as one of morphine.

Briggs smiled but declined politely when I tried to steer the conversation toward the potential merits of homeopathy. Based on her scientific credentials, I have to assume that's because she couldn't have had much good to say about the practice, but neither can she afford to enrage her advisory council, some of whose members believe in it wholeheartedly. Briggs has begun to push NCCAM, and the studies it funds, to focus more seriously on chronic pain. It's an excellent idea, since pain remains such a pervasive problem and conventional approaches at controlling it have shown only limited success.

Yet when an arm of the federal government devotes more than $100 million a year to a particular kind of research it makes a statement about the nation's health priorities. NCCAM has spent its money looking into everything from the use of qi gong as a way to treat cocaine addiction to Therapeutic Touch for bone cancer. "NCCAM is presented as a scientific vehicle to study alternative medicine's anomalous methods," Wallace Sampson wrote in "Why the NCCAM Should Be Defunded," the first of many appeals to close the center. "But it actually promotes the movement by assuming that false and implausible claims are legitimate things to study." As Sampson and many others have pointed out, while the center has demonstrated the "ineffectiveness of some methods that we knew did not work before NCCAM was formed," it has not proven the effectiveness of a single alternative approach to medical treatment.

That is because, uniquely among NIH centers, most of the research has been driven by faith rather than by science. D. Allan Bromley, a physicist who was science adviser to President George H. W. Bush, once said that NCCAM lent prestige to "highly dubious practices" that "more clearly resemble witchcraft than medicine." Paul Berg, a Stanford professor and Nobel laureate in chemistry, wrote that "quackery will always prey on the gullible and uninformed, but we should not provide it with cover from the N.I.H." And Ezekiel Emanuel, long the head of the department of bioethics at NIH, and brother of Rahm Emanuel, President Obama's chief of staff, has published widely on the ethics of placebo trials and the use of alternatives. Highly implausible or impossible methods, such as homeopathy, "psychic (distant) healing," Therapeutic—or Healing—Touch, and many other CAM claims are what Emanuel and colleagues have referred to as "trifling hypotheses," and the ethics of pursuing them are shaky at best. "Comparing relative value is integral to determinations of funding priorities when allocating limited funds among alternative research proposals," he wrote in the *Journal of the American Medical Association.*

Josephine Briggs and others at NCCAM may rely on the scientific method when assessing alternative therapies, but the man who founded the center certainly does not. In Senate testimony in March 2009, Harkin said that he was disappointed in the work of the center because it had disproved too many alternative therapies. "One of the purposes of this center was to investigate and validate alternative approaches. Quite frankly, I must say publicly that it has fallen short," Harkin said. The senator pointed out that since its inception in 1998, the focus of NCCAM has been "disproving things rather than seeking out and approving things."

That's a remarkable comment coming from a man who has spent years focusing on public health. In the opening sentence of the NIH mission statement, the agency is described as "the steward of medical and behavioral research for the Nation." It is not the steward for "validating" treatments that don't work. No wonder skeptics have complained from the beginning that Harkin created the center to promote alternative therapies, rather than to weigh their merit through rigorous testing. Afterward, I asked Briggs to comment on Harkin's statement. "I certainly understand the desire to see positive results," she said. "Rigorous, objective, scientific research often yields results other than what we would hope for, but our goal must remain one of building the evidence base regarding CAM practices."

Senator Harkin wasn't done. The next day, he traveled across town to the Institute of Medicine to address hundreds of people attending a "summit" on integrative medicine there. "Clearly, the time has come to 'think anew' and to 'disenthrall ourselves' from the dogmas and biases that have made our current health care system—based overwhelmingly on conventional medicine—in so many ways wasteful and dysfunctional," Harkin said. "It is time to end the discrimination against alternative health care practices."

Discrimination? The budget for NCCAM was $121 million in 2008, and it has disbursed more than $1 billion since Harkin first forced the Office of Alternative Medicine on the NIH leadership. By comparison, the NIH funding for autism research in 2008 totaled $118 million. Moreover, if a study is good enough to be funded, surely it can be funded by one of the other twenty-six institutes at the NIH. Physicians and other contributors to the relentless and insightful blog Science-Based Medicine have made this case many times. They have also examined in excruciating detail how NCCAM money is actually spent.

"Perusing the list of projects is truly depressing," wrote David Gorski of the Wayne State University School of Medicine. "True, a lot of the projects seem to be yet another study of Ginkgo Biloba, cranberry juice, or soy in various diseases. That's all well and good, but why is the study of natural products considered 'alternative' or 'complementary'? It's the same sort of stuff that pharmacologists have been doing for decades when they study most botanical products."

Gorski describes some of the weaker grants, including one funding a study called Polysomnography in Homeopathic Remedy Effects. "Yes, you have it right. Your tax dollars are going to fund at least a study this year on homeopathic remedies (a.k.a. water). But it's even worse than that. [One grant was] actually awarded to study homeopathic dilution and succussion"—the act of shaking liquid each time it is diluted—"and how they affect the dose-response curve of homeopathic remedies. I kid you not. I just about spit out my tea onto my laptop keyboard when I read it.

Naturally, it's at the Integrative Medicine Program at the University of Arizona" (which is run by Andrew Weil).

A FEW YEARS AGO I wrote an article for the *New Yorker* about a man named Nicholas Gonzalez. He was a highly credentialed physician, trained at Cornell University Medical School and at Memorial Sloan-Kettering Cancer Institute. Gonzalez worked out of an office in midtown Manhattan, where he was treating pancreatic cancer patients using some of the most amazingly bizarre methods in modern medical history. He prescribed twice-daily coffee enemas, for example, and a pill regime for most of his patients that ran to four single-spaced pages. (It included, in part, sixty freeze-dried porcine-pancreatic enzymes, capsules of adrenal medulla, amino acids, bone marrow, selenium 50, thyroid, vitamin A 10,000, and vitamin E succinate. And many, many more.)

The prognosis for pancreatic cancer patients, then and now, is particularly bleak, and it seemed to me at least that if one were going to roll the dice, it might make more sense with that particular disease than with another. Gonzalez was reviled by members of his profession. But unlike virtually any other alternative healer, he always insisted that he wanted his method to be tested in the most rigorous possible way by the NIH, and in 2000 he received a $1.4 million grant to do just that.

Most people in the medical world thought the trial was a waste of time and money. I didn't agree at the time, but I do now. Studies like that just make the ridiculous seem worth investigating.

Wallace Sampson was right—we cannot afford to fund research that has no reasonable chance of success. It wastes money and steals time that could be devoted to more promising work. More than that, though, it makes the denial of reality acceptable. If you believe that coffee enemas and energy fields offer hope, you can believe anything. We have seen where that can lead.

Since the beginning of the AIDS epidemic, a small band of committed activists led by Peter Duesberg, a microbiologist at the University of California at Berkeley, have denied that the epidemic is caused by the human immunodeficiency virus. Thousands of molecular studies (and millions of deaths) have made it clear that there is no possibility they are right, but with the help of the Internet, the impact of that denialism has been felt throughout the world. People want to be told that everything is going to be all right. That's normal. ("What mattered to me as a person living with HIV was to be told that HIV did not cause AIDS. That was nice," Winstone Zulu, a Zambian AIDS activist and former denialist, has written. "Of course, it was like printing money when the economy is not doing well. Or pissing in your pants when the weather is too cold. Comforting for a while but disastrous in the long run.")

AIDS denialism doesn't die, even in America. In the United States the group the Foo Fighters recorded the soundtrack for a documentary called *The Other Side of AIDS,* which was directed by the husband of Christine Maggiore, one of America's more prominent AIDS denialists (who, in 2008, died of AIDS). And in 2009, *House of Numbers,* a film made by another AIDS denialist, Brent W. Leung, provoked new outrage (and letters from nearly every prominent scientist featured in the film saying they were misled and quoted out of context).

It is of course in South Africa that the denial of fact-based medicine has had its most deadly effect. South Africa has the world's largest population of people infected with HIV. Instead of treating those people with the antiretroviral drugs necessary to save their lives, former president Thabo Mbeki denied for years that the virus caused the disease. Like the Zambian leaders who refused to accept genetically engineered food to feed their starving people, and the northern Nigerian mullahs who campaigned against polio vaccinations, Mbeki suspected a Western plot. In this case, he believed that Western pharmaceutical companies had banded together to threaten the future of Africans; and he was convinced that a natural, local solution would be far more effective than the "poison" offered by such organizations as the World Heath Organization and UNAIDS.

Instead, he and his longtime health minister Manto Tshabalala-Msimang recommended herbs, garlic, and lemon. I have seen the effect of those herbs, and of the vitamin regimens peddled by people like the German health entrepreneur Matthias Rath with the tacit support of Mbeki and Tshabalala-Msimang. Rath urged people to substitute remarkably high doses of multivitamins for proven AIDS therapies like AZT. People who did that, rather than relying on the antiviral medicines they needed, died. In 2008, a group of researchers from the Harvard School of Public Health concluded that the South African government would have prevented the premature deaths of as many as 365,000 people between 2000 and 2005 had it provided antiretroviral drugs to AIDS patients. The study also concluded that the drugs were largely withheld due to "Mbeki's denial of the well-established scientific consensus about the viral cause of AIDS and the essential role of antiretroviral drugs in treating it."

Taking megadoses of vitamins or craniofacial massage for the flu may seem comforting. At worst, many have argued, such actions are self-inflicted wounds—like the self-inflicted wound of refusing to vaccinate a child. There comes a point, though, when individual actions become part of something bigger. Progress is never guaranteed. It can vanish if reality ceases to make more sense than magic. Denialism is a virus, and viruses are contagious.

5

Race and the
Language of Life

In the spring of 1998, a team of researchers from the Centers for Disease Control traveled to a meeting of the American Thoracic Society in Chicago, where they presented a report on the severity of asthma among Hispanics. Minorities living in America's largest cities visit the emergency room more often, spend more time in the hospital, and die in far greater numbers from asthma than the rest of the population; they are also far more likely to develop pneumonia and other pulmonary diseases. None of that was news to most of those who attended the meeting, but the CDC study was the first to focus specifically on the prevalence of asthma in Hispanics. During the course of his presentation, the pulmonary specialist David Homa pointed out that he and his colleagues had run across one particularly surprising result in their research: Hispanics living in the Northeast of the United States were three times more likely to develop asthma than Hispanics in the South, Southwest, or West.

Many people in the audience found that odd; differences in the rates of pulmonary disease are often a result of social conditions, the environment, and disparities in the quality of health care. Poor people rarely receive the best possible treatments and consequently they don't do as well as richer patients—with asthma or with most other illnesses. The study was designed to account for those facts. Even so, a Hispanic man in New York was more likely to get sick than one in Los Angeles or Chicago. Some participants shook their heads in surprise when they saw the data, but Esteban González Burchard was not among them. Burchard, at the time a twenty-eight-year-old internal medicine resident at Boston's Brigham and Women's Hospital, had known for years that he wanted to specialize in pulmonary disease, largely because of its punishing effect on minorities. He sat riveted by Homa's presentation, and particularly by the data that suggested the illness seemed so much worse on the East Coast than in other parts of the United States.

"I jumped when I heard him say that," Burchard told me when we first met in his office at the University of California at San Francisco, where he is assistant professor in the departments of biopharmaceutical sciences and medicine. "I am Hispanic and I have lived on both coasts and I knew that the obvious difference there had to be between Puerto Ricans"—who reside principally in the East—"and Mexicans"—who are more likely to live in the West. At the time, Burchard was working in the laboratory of Jeffrey Drazen, a professor at the Harvard School of Public Health, who was soon to become editor of the *New England Journal of Medicine*. In his lab, Drazen had identified a genetic risk factor that would explain the differences in asthma severity between African Americans and Caucasians. Both he and Burchard thought the CDC data might help explain genetic differences within the Hispanic community as well.

"I talked to David Homa and suggested that this data could very possibly be the result of genetics," Burchard said. "I don't know if he thought I was a little crazy or what." After all, genetics seemed unlikely to provide an explanation for such a striking disparity among people with a common ethnic heritage. The prevailing view since the early days of the Human Genome Project has been that such differences no longer seem worth thinking about, and many notable researchers have argued that focusing on race in this way is not only scientifically unsound but socially dangerous. Yet something had to account for the wide gap, so Burchard persuaded Homa that the CDC ought to take a closer look at the data. Two years later, the CDC study, now focusing on the differences among Hispanics of Cuban, Puerto Rican, and Mexican descent, was published. It showed that the prevalence, morbidity, and mortality rates of people with asthma varied significantly within those groups and concluded that genetics seemed to be at least partly responsible.

Burchard, like most physicians, believes that social and economic disadvantages explain much about why minorities in the United States suffer disproportionately from so many diseases— eight times the rate of tuberculosis as whites, for example, ten times the rate of kidney failure, and more than twice the rate of prostate cancer. Tuberculosis has served for generations as the signature disease of urban crowding, homelessness, and poverty, and many people who die of AIDS do so because the virus makes them susceptible to infections that cause pneumonia. But Burchard asked himself whether economics and environment alone could explain why one group of Latinos (Puerto Ricans) had among the highest rates of asthma in the United States while another group (Mexicans) had virtually the lowest. That made no sense. "I was convinced then, and am even more convinced now," he told me, that "there are specific ethnic, genetic, and environmental risk factors in play here." For the past decade, Burchard has worked with one principal goal in mind: to understand the meaning of such genetic differences between racial groups. It has not been easy, nor has the research always been particularly welcome. Grant money has often been hard to find, and skepticism from colleagues palpable. But Burchard persisted.

In 2001, with the help of senior scientists from medical centers in the United States, Mexico, and Puerto Rico, he embarked on an exhaustive attempt to better understand the clinical, genetic, and environmental differences in severity among Mexican and Puerto Rican asthmatics. The investigative study, called GALA—Genetics of Asthma in Latino Americans—has not only revealed genetic differences within Hispanic populations. Perhaps more importantly, it has demonstrated that despite having more severe asthma, Puerto Ricans respond less well to the standard treatments they are most likely to receive. Albuterol, for example, is the most commonly prescribed asthma medication in the world, and frequently the sole drug people receive (or need) to treat their asthma. Yet for many Hispanics it can prove useless. Nonetheless, albuterol is often the only asthma drug prescribed for Puerto Ricans or Mexicans in the United States.

"The idea has become fashionable that we are all one species and that ethnicity and race do not play defining roles in determining the causes of disease," Burchard said. "But look at the data. The one-size-fits-all approach to medicine and to drug therapy does not work. We see that over and over again. I can't think of anything more important in medicine right now than trying to tease out the causes of these differences. But believe me, there are many people who think I am wrong." He sat back, shook his head, and smiled darkly. "There are serious scientists who say we should not even do this kind of research, that races should be treated as one and that the genetics of humanity are not diverse enough to play this kind of role in diseases. You can't look at the data and make those assumptions," Burchard said. "But if reality upsets people, they will simply look in another direction. People deny what makes them uncomfortable, and many—even in my business—say we shouldn't use the word 'race' at all."

It has never been easy to invoke the subject of race in America. Discrimination has long been as obvious in medicine as in other areas of society. In the era of personalized medicine, where relevant new information seems to appear daily, the issue has become more volatile than ever. At many meetings where race and genetics are discussed, researchers spend as much time debating semantics as they do discussing scientific results. (In 2008, one participant at a National Institutes of Health conference on genomics and race argued that not only is the term "race" unacceptable, but so is the term "Caucasian," because it implies racial rather than geographic ancestry.)

"How much of a factor does genetics play in these research results?" Burchard continued. "I will tell you honestly, I don't know— but neither does anyone else. I am Hispanic and I want people to get the treatment that serves them best. I want every tool at my disposal and every tool at the disposal of my patients. Genetics is one of the most powerful weapons we have. Yes, we are strikingly similar in many—even most—ways. But genetics also makes us different. That scares some people, but it is a fact.

"We are finally coming to a period in scientific history where people may be able to benefit from those differences. Where they can actually help treat and cure diseases. You would think that is something that scientists would support. But too often it is not. Let's face it, in this country there have been major efforts, guided by endless waves of political correctness, to close the door to the possibility that there could be important racial differences among human beings. At first I found it surprising. I don't anymore. But be honest: who can possibly benefit from this approach to medicine?"

ON JUNE 26, 2000, surely one of the most notable days in the history of science, President Bill Clinton announced the completion of the first draft of the Human Genome Project. He spoke in the East Room of the White House, where Thomas Jefferson and Meriwether Lewis had presented the map of Lewis and Clark's historic expedition to the public. The symbolism was impossible to ignore, as were the parallels between what Clinton described as Clark's "courageous expedition across the American frontier" two centuries earlier and the Human Genome Project's exploration of the contours and complexities of the human cell. Many of the scientists who had struggled to compile that blueprint of human DNA—the string of three billion pairs of chemical "letters" that make up our genetic code—stood by the president's side. "Without a doubt, this is the most important, most wondrous map ever produced by humankind," Clinton said. "We are learning the language in which God created life." British prime minister Tony Blair, joined the conference by satellite, along with researchers who had contributed to the work in England. Science, Clinton said, was on the verge of gaining immense power to heal, power that until recently we could not have imagined. He noted that the information packed into the structure of our genome was so valuable, and the potential benefits of understanding it so great, that it was conceivable that "our children's children will know the term cancer only as a constellation of stars."

That's unlikely, but Clinton's optimism hardly seemed misplaced, and it seems even less so now. By assembling a complete map of the human genome, and then refining it literally every day, geneticists have already transformed fields as diverse as anthropology, history, molecular biology, and virology.

An entire industry, genomics, has emerged to study the structures and functions of genes and how they interact with each other. The hereditary information contained within our genes, our DNA, is written in a four-letter language that, if printed out, would fill more than a thousand New York City telephone books. (Each letter corresponds to one of four nucleotide bases: A for adenine, T for thymine, C for cytosine, and G for guanine.) These sequences, arranged in millions of threadlike helixes and passed from one generation to the next, carry within them the instructions required to assemble all living things—a set of instructions which genomic scientists are working feverishly to decode.

The resemblance among humans is startling: compare one person with any other, chosen randomly from any two places on earth, and genetically they will be more than 99 percent identical. It doesn't matter whether one of those people is from Sweden and the other from Zambia, or whether they are twins, or of different genders. Yet, there are still millions of places in our genome where that recipe varies among individuals by just a single genetic letter. Those places are called single-nucleotide polymorphisms, or SNPs (pronounced "snips"). SNPs are useful markers for different versions of genes, and they help emphasize the differences that scientists are trying—with increasing success—to associate with various diseases. Over the past decade, researchers have been able to use these surrogates as molecular guideposts to identify scores of genes that play major roles in diseases ranging from prostate cancer to age-related macular degeneration.

The Human Genome Project launched a modern Klondike, and billions of dollars have been invested in an attempt to understand the exact structure of virtually every gene in the human body and then translate that knowledge into effective drugs. We are in the earliest stages of this vast effort at sifting through the raw data that contains the language of life. Eventually, however, genomics will almost certainly provide the information necessary to help answer many of the most fundamental questions we can ask about ourselves and about biology. To what extent are genes responsible for how we grow, think, evolve, become sick, and die? Are traits that pass through generations in a community genetically determined, or are they the expression of cultures that have been shared for thousands of years? Is it even possible to quantify how much of what we are comes from genes and how much from the circumstances of our lives?

The last question is the most important because when we understand how humans are put together we will have a much better grasp of the genetic basis of major diseases—what causes them and what causes people to vary so dramatically in their ability to respond to particular medicines. As every physician knows, drugs that work well for one person don't necessarily work for others. Some, like albuterol, are effective on whites but not on many Hispanics. For African Americans, who suffer from congestive heart failure at twice the rate of whites, there is BiDil—a combination of two older drugs, which when taken in concert turn out to work far better for blacks than for whites. (The drug, when used in this fashion, became the first race-based medicine approved by the FDA, specifically to treat blacks. The action caused intense controversy, but it also offered a new form of relief to African Americans with heart failure.)

While vital clues to causes of various afflictions emerge in a torrent, they rarely provide definitive answers. But the initial map did prove conclusively that all humans share a nearly identical genetic heritage. Many researchers have even argued that relying on race as a way to define and connect large groups of seemingly similar people no longer makes sense, except as a way to discriminate against them.

President Clinton made a point to stress that at his news conference. After all, what could be more exciting to a liberal politician raised in the South than unshakable evidence that racism was based on a series of socially created misconceptions about human evolution? As Eric Lander, a genomic pioneer and the director of the Broad Institute, the research collaboration between Harvard and MIT, put it, "Racial and ethnic differences are all indeed only skin deep."

J. Craig Venter, who at the time was president of Celera Genomics, the private company that dueled with the government to complete the project first, also attended the ceremony. There is no more compelling or astute scientist in the world of genomics than Venter, who in 1992 founded the Institute for Genomic Research. A comment he made that day, that "the concept of race has no genetic or scientific basis," has been repeated often. So have the remarks of his main competitor, Francis S. Collins, the renowned geneticist who led the federal effort to map the genome, pointed out that the data showed there were probably more significant genetic differences *within* racial groups than between them. (In 2009, Collins was named by Barack Obama as director of the National Institutes of Health.) All those comments provide support for the comforting idea of a family of man, even a society without race.

Nor should such findings be surprising; we are a young species, one that migrated out of Africa and throughout the world only about one hundred thousand years ago—relatively recently in evolutionary terms. Many scientists already understood this well. In fact, during the first decade of the Human Genome Project some participants were so convinced of the homogeneity of humanity that they insisted that the genomic sequence of *any one person* could be used as a basic reference template for everyone else on earth.

There was a widely shared feeling, according to M. Anne Spence, who was on the Genome Project's ethics committee, that the sequences would be the same and that gender ought not to matter. Spence and several others insisted that such thinking made no sense. For one thing, men carry a Y chromosome and women two Xs. That would have to account for *some* differences. Nonetheless, most drug research in America has been carried out on middle-aged white men. People often have radically different responses to the same medicines—and women, in particular, react in ways that men do not. Spence pointed that out and also argued that sequencing the genome was not simply a scientific enterprise, but one with lasting implications for our political, social, and cultural lives.

She may have been more right than she knew. In its early days, instead of settling debates about race and medicine, the Genome Project inflamed them. By the time the genome was published, nearly a year after Clinton's announcement, two distinct camps had formed: those who believed race no longer existed as a biological entity, and those who argued that race and ethnic background continued to provide crucial information for medical research. Debates erupted in scientific journals, at academic conferences, on university campuses, and even within the federal government's scientific establishment. Considering the term's origins, the anxiety is not hard to understand. Classifying people by race has had a profoundly disturbing history, leaving a legacy of hatred throughout the world. Naturally, then, many scientists rushed to embrace our remarkable genetic similarities as a way to dismiss race entirely. As Robert Schwartz of Tufts University argued in a widely circulated article published in the *New England Journal of Medicine,* "Race is a social construct not a scientific classification."

He went on to point out that racial identification plays its most important—and destructive—role in setting social policies. In one way or another race has been used to justify much of the abuse humans have inflicted on each other throughout history. Putting race together with medicine has been particularly explosive. One has only to think of the Tuskegee Experiment to see that. Race has been used to justify eugenics, and more than once to justify genocide. The facts of the human genome suggested that it might be possible to move beyond such divisive ways of thinking about our species. "Sadly," Schwartz wrote, "the idea of race remains ingrained in clinical medicine. On ward rounds it is routine to refer to a patient as black, white, or Hispanic yet these vague epithets lack medical relevance."

Most scientists think it will be years, possibly decades, before we reap the full intellectual harvest of the Human Genome Project. After comprehensive study, we can explain a small percentage of genetic links to common disease. But there is much more we don't understand—including how some genes work to protect us from illnesses that other genes cause. Meanwhile, that 99 percent figure has been published everywhere and is used as the basis of a propaganda war by both sides in the race debate. There is no disputing our homogeneity. It is also true, however, that we share 98.4 percent of our genes with chimpanzees. Few people would argue that makes us nearly identical to them. Even drosophila—the common fruit fly—has a genetic structure that shares almost two-thirds of its DNA with humans. Does that mean we are *mostly* like fruit flies? The simple and largely unanswered question remains: what can we learn from the other 1 percent (or less) of our genome that sets us apart from everyone else?

"WHAT WE ARE going to find is precisely that the other percent plays a role in determining why one person gets schizophrenia or diabetes while another doesn't, why one person responds well to a drug while another can't tolerate it," Neil Risch said. Risch, who is Lamond Distinguished Professor and director of the Institute for Human Genetics at the University of California at San Francisco, argues that the concept of race remains highly valuable in medicine, and that people only pretend otherwise out of a misguided sense of decency. "It's crazy to banish race just because it makes people uncomfortable," he said. "It's a genuinely nice idea and I understand the reasons for it. But scientifically it just makes no sense.

"These are imperfect but valuable ways to describe a group," he continued. "You can talk about age in the same way. Rarely would a person's chronological age correspond exactly with his biological age—for both environmental and genetic reasons. Using a birth year is not necessarily a precise way to measure it. We all know there is ageism in society. Does that mean as physicians we should ignore a person's date of birth? Of course not. It's an important tool in our arsenal."

Risch is one of the most prominent and highly respected geneticists in the United States, and when it comes to this issue, one of the most controversial. He is also antsy; when we met in his office he would sit quietly for a few moments, then burst into speech and then abruptly stop. "Here's the deal," he said, "especially for complex major diseases. You see differences in health rates and is your first conclusion going to be that they are based on genetics? Of course not. Inevitably, the environment is playing some role and the interaction between environment and genetics is incredibly complex.

"If you go to morbidity and mortality statistics, what do you find?" he continued. "You find that African Americans generally have higher rates of disease and death across the board for everything; all cancers, heart disease, just about everything. This just doesn't make sense genetically. The differentiation between racial groups is not big enough for one group to have all the genes for disease. So of course it's environmental." He smiled, pausing for emphasis. "But when you eliminate the environmental differences you are *still* left with a significant disparity between races, and there is where genetic factors may play more of a role."

Risch rattled off a list of diseases in which genetic variations between ethnic groups had been observed: Crohn's disease is more common in people of European heritage, and Risch's team identified a SNP that confers a much higher risk on Europeans than on any other geographical group. "This is clear and unequivocal," he said. "Those SNPs don't exist in Asians or in Africans. There are others. Hemochromatosis"—a condition in which the body produces and stores too much iron. "Between 8 and 10 percent of Europeans have this mutation. In the rest of the world, though, it is almost nonexistent." Perhaps the most interesting example is of a particular protein used by the AIDS virus to dock with cells and infect them. The Delta 32 mutation on the CCR5 receptor prevents that; the virus can't find a convenient way to lock on and infect cells that carry this mutation, which is present in as many as 25 percent of white people, particularly in northern Europe. The mutation has never been found in Africans or Asians.

"There are real, powerful, and useful implications to all this," Risch said. "Interferon is a treatment used for hepatitis C. Forty percent of Caucasians respond well to it and actually clear the virus out of their system. Africans don't respond at all. Not at all. This matters immensely. It's not socio-cultural or economic. It seems to be genetic. And we need to know this, because giving blacks interferon when they have hepatitis C is not going to help them. We have to come up with other treatments."

Late in 2008, the National Human Genome Research Institute held a forum at which genetic researchers discussed with ethicists how best to present their discoveries to the public. Studies that underscore racial differences are almost always in dispute. In 2005, the geneticist Bruce Lahn and colleagues at the University of Chicago published two papers that described their investigations into the evolution of the human brain. Lahn found that mutations in two genes that regulate brain development were more common in Eurasians than in Africans. That implies that those variants conferred a survival or reproductive benefit, and that they emerged after humans left Africa.

Nobody knows what those genes do, and there was no evidence to show that they had acted on intelligence. Nonetheless, putting the words "gene," "brain," and "race" together in a sentence is bound to cause trouble. People on both sides of the political divide leapt to conclusions; Lahn, a lifelong liberal, was embraced by the right and denounced as a sensationalist even by some of his colleagues. He had stressed that the study had no racial component per se, and that genes other than those in the brain could have caused their selection. Nor is it clear what, if anything, those mutations represent. But because they were less common among sub-Saharan Africans than in other populations, the work caused a sensation. Still, Lahn has wondered for years whether there might be a genetic element to variations in social status. "You can't deny that people are different at the level of their genes," he said at the time, citing the examples of skin color and physical appearance. "This is not to deny the role of culture, but there may be a biological basis for differences above and beyond culture."

That kind of talk infuriated his colleagues, and it still does. At the NIH workshop, Celeste Condit, a professor of speech communication at the University of Georgia, spoke about the way she thought Lahn's study was framed. "The papers could be seen as having a political message," Condit told *Science* magazine: in other words, the research might have implied that those genes contribute to differences in IQ. Lahn, who has since shifted the focus of his work to stem cell research in part because of the controversy, has repeatedly stated he had not meant to suggest that.

During the bicentennial celebration of Darwin's birth, in 2008, the journal *Nature* invited distinguished scientists to debate whether the subject of race and IQ was even worthy of study. The dispute was lively. "When scientists are silenced by colleagues, administrators, editors and funders who think that simply asking certain questions is inappropriate, the process begins to resemble religion rather than science," Stephen Ceci and Wendy M. Williams, geneticists at Cornell University, wrote. "Under such a regime, we risk losing a generation of desperately needed research." The British neuroscientist Steven Rose disagreed fundementally, calling the study of the relationship between race and IQ "ideology masquerading as science."

Despite the subject's volatility, and the fact that most people would prefer to deny its implications, neither the federal government nor the pharmaceutical industry is quite ready to abandon the concept of race. In March 2008, the National Institutes of Health announced the establishment of the Center for Genomics and Health Disparities. (If there were no genomic disparities, why establish such a center?) A few months earlier, the Pharmaceutical Research and Manufacturers of America had released a lengthy report describing nearly seven hundred new drugs that were under development to treat diseases that disproportionately affect African Americans. (There was more than a little marketing behind the report; many of those drugs, should they make it through the FDA approval process, would also prove beneficial for other ethnic groups.)

Some of the genetic factors involved in drug response have been known for decades and can be attributed to proteins called drug metabolizing enzymes. Differences in the genes that encode these molecules are responsible for how quickly the enzymes process and eliminate drugs from our bodies, as well as how they are broken down in the blood. If a drug is metabolized too quickly, it may not reach a high enough concentration to work properly. If it is metabolized too slowly, however, enough of that drug could accumulate to reach a toxic level in the body. In either case, the patient would suffer, but none of that is news to physicians with ethnically diverse patient populations.

"Almost every day at the Washington drug clinic where I work as a psychiatrist, race plays a useful diagnostic role," Sally Satel wrote in a much-debated 2002 New York Times article entitled "I Am a Racially Profiling Doctor." She has written often about the subject. "When I prescribe Prozac to a patient who is African-

American, I start at a lower dose, 5 or 10 milligrams instead of the usual 10-to-20 milligram dose. I do this in part because clinical experience and pharmacological research show that blacks metabolize antidepressants more slowly than Caucasians and Asians. As a result, levels of the medication can build up and make side effects more likely. To be sure, not every African-American is a slow metabolizer of antidepressants; only 40 percent are. But the risk of provoking side effects like nausea, insomnia or fuzzy-headedness in a depressed person—someone already terribly demoralized who may have been reluctant to take medication in the first place—is to worsen the patient's distress and increase the chances that he will flush the pills down the toilet. So I start all black patients with a lower dose, then take it from there."

The main argument against relying on race in this way is simple but powerful: different races receive decidedly different standards of health care in the United States, and that is unacceptable. The disparity explains why African Americans and Hispanics have more chronic illnesses than whites, and why they take longer to recover from them. Genetics is only one piece of a puzzle; if we place too much emphasis on it we will invariably continue to neglect more significant reasons for the gulf that separates the health of black and white Americans. You don't have to be an astute student of the United States, or the history of the modern world, to take such concerns seriously.

Even so, respect for other ethnic groups cannot alter biological reality. Everyone knows how different we all are from each other. Some of us are dark and some are light, some tall and others short. It's genetic; we inherit those traits from our parents. In fact, entire industries cater to differences like these: nobody would expect to see Barack Obama wearing the same size suit as Kobe Bryant.

"Many of my colleagues argue that we should banish the word 'race' completely," Neil Risch told me. "They say let's use different words. Instead of race we should talk about geographic distribution of ancestors. And that's completely fine with me; we can call it GOAD. Now, think about that for two minutes, and then tell me: if we described people that way, do you actually believe there would be no 'goadists'?"

ICELAND MIGHT SEEM like an odd place to search for answers to complex questions about race and genetics. The country has three hundred thousand residents, all of whom are so genetically similar that telephone numbers are organized by first names in the Reykjavik phone book. A thousand years of volcanic eruptions and other catastrophes have had the effect Darwin would have anticipated: those plagues and natural disasters pruned the population and cut back sharply on the genetic diversity of the island. As a result, the hereditary instructions of the entire nation have passed through a small gene pool for fifty generations.

There are thousands of illnesses—like cystic fibrosis, sickle-cell anemia, and Huntington's chorea—whose cause can be traced directly to the mutation in a single gene. They usually follow simple Mendelian patterns of inheritance and run in families. Most major diseases, on the other hand, including cancer and cardiovascular illnesses, which kill millions of people every year, are the result of a complex combination of environmental history,

behavioral patterns, and the interaction of hundreds of genes working together in ways that even now we only dimly understand.

The most direct approach to finding the origins of those
diseases is to compare the DNA of people who are sick to the
DNA of their healthy relatives (and ancestors). When a group is
almost identical, their differences become much more apparent.
Those kinds of studies are hard to conduct in a racially and ethnically diverse country like the United States, where ancestors can
rarely be traced for more than a few generations. If one group's
cultural heritage, environment, and habits differ from another's
then so almost certainly are the causes of its illnesses. That's not a
problem in Iceland. Despite centuries of seclusion on a remote
island in the North Atlantic, people there develop serious diseases
at roughly the same rate as people in other industrialized countries. There is no place more ideally suited for research into the
genetics of major diseases.

"What do race and genetics have to do with common diseases?"
bellowed Kari Stefansson when I asked to discuss the subject with
him. He looked as if his eyes were ready to burst. "Everything,
obviously. How can you be stupid enough to ask that question?"
We were standing in his office at deCODE genetics, the company
he founded in 1996 to mine the genetic heritage of the Icelandic
people. Stefansson is six feet five inches tall, dresses almost exclusively in black, and is famously imperious. When he hovers over
you and calls you an idiot it makes an impression that doesn't fade
quickly. The first time we met, nearly a decade ago, I couldn't
believe that Stefansson could be so condescending. Since then I
have come to regard his conversational manner as a personal

trait, like freckles or a twitch. Throughout that time, Stefansson's self-confidence has never wavered—and that's not wholly without reason.

Perhaps more than any scientific institution other than the U.S. National Institutes of Health, which is funded by the federal government, deCODE is responsible for producing a stream of genetic information that promises to change medicine in ways that even a decade ago would not have seemed possible. Almost no day passes without some revelation describing how our genes influence the way we live, behave, get sick, and die. DeCODE has isolated genes that are associated with type 2 diabetes, prostate cancer, heart attack, obesity, and schizophrenia, to name just a few. The company has even unearthed tentative hints at the relationship between fertility and longevity. All by homing in on differences in the DNA of people who are as alike as any group on earth.

"Differences matter," Stefansson said, striding into his office with a protein drink in each hand. "They matter enough to cure diseases and save million of lives. Race. Geographical ancestry. Call it what you want. If our work has shown us anything, it ought to be that the even smallest of goddamn differences matter." Stefansson spent more than a decade at the University of Chicago, where he became a tenured professor of neurology. He returned to Iceland briefly in the early 1990s to run the Institute of Pathology, then the country's most distinguished scientific research organization. He was restless, though, and for five years moved back to the United States as a professor of neurology and pathology at the Harvard Medical School. It was then, during a brief visit home to conduct research on his specialty, multiple sclerosis, that Stefansson realized Iceland was a genetic jackpot.

The deCODE building, just a brief walk from the center of old Reykjavik, is crafted from the stark school of Nordic realism, all plate glass and angular bits of steel. It is eerily clean and quiet and the mood seems surreal: perhaps that's because I have only visited in the middle of winter, when the sun sets before noon, and at the height of summer, when people play chess in the courtyards until four a.m. Despite its unparalleled research success, the company has been badly hurt by Iceland's economic shipwreck, not to mention some unlucky investment decisions and its own outsized ambitions. DeCODE never saw itself solely as a research center. It intended to become a major biotechnology and pharmaceutical company, but those plans have largely remained unfulfilled.

Nonetheless, deCODE helped start a revolution. Fueled by the almost unimaginably rapid growth in sequencing power, genomics is beginning to transform the way we think about medicine, and about the rest of our lives. The benefits, particularly drug treatments tailored to individual needs, have been overly hyped, as new technologies always are. In the past, it often took twenty-five years to turn a scientific discovery into a common therapy. (Or longer. The German chemist Adolf Windaus won the Nobel Prize in 1928 for work that helped determine the chemical composition of cholesterol. It took almost a century until that discovery made its way into a class of drugs—statins—now taken by millions of people every day.) Powerful computers and gene sequencing technology are changing all that, supplying the vocabulary necessary to make sense of the digital information contained within each of our bodies—and each of our cells.

SNPS PROVIDE A useful way to calculate a person's genetic risks of developing scores of diseases. Yet they are a half-measure, an imperfect substitute for the information that comes from scanning an entire genome—which still costs $100,000. The price won't stay high for long. (In fact, one company, Complete Genomics, claims it will be able to sequence an entire human genome for $5,000 by 2010.) The cost of combing through billions of bits of DNA has fallen by a factor of more than one hundred thousand in less than two decades. In 1990, as the Human Genome Project got under way, scientists estimated that sequencing a single genome would cost $3 billion. The final bill is hard to calculate, because the figures include the cost of many scientific activities relating to genomics carried out during the thirteen-year-long project. But the total was far less than the original estimate, and when the project ended in 2001, the team said that they could do it again for $50 million.

Five years later, the molecular geneticist George Church said he could sequence a genome for about $2 million. The following year, it took two months and less than $1 million to sequence the complete genome of James Watson, who in 1953 discovered the structure of DNA along with Francis Crick. A drop in cost from $3 billion to $100,000 in twenty years is impressive. Time is an even more useful measuring stick: what took thirteen years in 1988 and two months in 2007 will almost certainly take less than five minutes within the next two or three years. Church, who is director of the Lipper Center for Computational Genetics at Harvard Medical School, and holds dual positions at Harvard and MIT, expects to see steeper price declines and the faster sequencing rates that come with them, soon. Church helped develop the earliest sequencing methods, nearly twenty-five years ago, while working in the lab of the Nobel Prize–winning chemist Walter Gilbert.

"I don't know whether we can squeeze it down by a factor of one hundred in the next year or so—it's hard to even guess what the cost will be in five years. But it will be low," he said. "You just don't get that kind of change in any other industry." In 2007, Church embarked on his most audacious undertaking, the Personal Genome Project. He intends to sequence the genomes of one hundred thousand volunteers—he has already sequenced and published the genomes of the first ten. The eventual database will prove invaluable in correlating genomic information with physical characteristics. Researchers will have access to the database at no cost. Naturally, without the rapid evolution of sequencing technology the project would not have been possible.

"In 1984, thirty base pairs"—thirty rungs on the helical ladder of six billion nucleotides that make up our DNA—"was a good month's work," Church told me. "Now it takes less than a second." Craig Venter, who knows as much about how to sequence a genome as anyone, agrees. "I spent ten years searching for just one gene," he said. "Today anyone can do it in fifteen seconds." Indeed, the X Prize Foundation has offered $10 million to the first group that can sequence one hundred human genomes in ten days at a cost of $10,000 or less per genome. As many as two dozen teams are expected to compete.

In 2007, seizing on the cascade of genetic information that had suddenly become acessible, deCODE and two California companies, 23andme and Navigenics, began to sell gene-testing services directly to consumers. The tests analyze up to one million of the most common SNPs—a small fraction of our genome—focusing on the most powerfully documented relationships between those SNPs and common diseases. For each disease or condition, the companies estimate the risk of a healthy person developing that illness. Both deCODE and 23andme sold their first tests for just under $1,000, but prices keep falling. By the end of 2008, a 23andme test cost $400. Navigenics charges $2,500 for its full regimen, which includes the services of genetics counselors; deCODE offers packages at various prices.

Much of deCODE's research relies on its own formidable database, while 23andme, whose slogan is "Genetics just got personal," has emphasized genealogy and intellectual adventure, not just medicine, and encourages customers to share data, participate in research studies, and form social networks on its Web site. In 2008, *Time* magazine named the 23andme test as its invention of the year, but critics have described the company's approach as frivolous because it not only provides disease information but also helps customers learn about less useful—but perhaps more amusing—traits like whether they have dry ear wax or can taste bitter foods. Nobody disputes the quality of the company's science, however, or its standards. (I should state clearly, and for the record, that the founders of 23andme are close friends of mine, and have been for years.)

The testing process is similar at each company. After spitting into a tube or swabbing their cheeks for saliva, customers submit samples of their DNA. Within weeks, they receive an e-mail informing them that they can retrieve their information from a secure Web site. These are not diagnostic tests and their predictive value is subject to much debate. Many diseases involve the interaction of scores or even hundreds of genes. A SNP that shows a heightened risk for a particular condition almost always only tells part of the story, and some people worry that since the data is rarely definitive customers might be misled. "We are still too early in the cycle of discovery for most tests that are based on newly discovered associations to provide stable estimates of genetic risk for many diseases," wrote Peter Kraft and David J. Hunter, both epidemiologists at the Harvard School of Public Health, in an article titled "Genetic Risk Prediction—Are We There Yet?" in the April 16, 2009, issue of the *New England Journal of Medicine.* They thought not. "Although the major findings are highly unlikely to be false positives, the identified variants do not contribute more than a small fraction of the inherited predisposition."

None of the services pretend that genetic tests alone can explain complex health problems. On its Web site, 23andme states that "in order to make a diagnosis, your doctor considers not only your genetic information, but also your particular personal and family history and your physical condition, as well as any symptoms you are experiencing. Your genotype is only part of the equation." Making a similar point, deCODE suggests that you explore your genetic risk factors and keep a vigilant eye on your prospects for prolonged health. Even Navigenics, the most clinically oriented of the three, tells prospective customers that there are no certain answers in the information they provide: "This level of personalization may help you take action to detect health conditions early, reduce their effects or prevent them entirely."

Even knowledgeable consumers can struggle to put partial

genomic data into perspective, particularly if a report indicates that they are at greatly increased risk of developing a serious illness. That information is based on *what is known*—which in most cases is only a fraction of what there is to learn. Three prominent health officials, including the editor of the *New England Journal of Medicine* and the director of the National Office of Public Health Genomics at the Centers for Disease Control, have suggested that until better data is widely available, a person would do more to improve his health outlook by "spending their money on a gym membership or a personal trainer."

Caveats and caution are necessary because risk is relative and few people deal with abstract probabilities rationally. If, for instance, a person has four times the normal risk of developing a particular disease, should he worry? That is an extremely elevated figure. Without context, however, a number means nothing. Take the digestive disorder celiac disease; fewer than one in a hundred people develop celiac disease in the United States, so a relative risk figure four times the average would mean that you stand a slightly greater than 96 percent chance of avoiding it completely.

That doesn't mean genomic tests aren't useful. They can change (and save) your life. Jeff Gulcher, deCODE's forty-eight-year-old chief scientific officer, wasn't around the last time I was in Reykjavik. Stefansson and he have worked together for more than two decades, since the day that, as a graduate student, Gulcher walked into Stefansson's laboratory at the University of Chicago.

When Gulcher took his deCODEme test, a month before I arrived, he learned that his relative risk of developing prostate cancer was 1.88. That meant he was almost twice as likely as the average person to get the disease. Gulcher took those results to his physician, who ordered a prostate-specific antigen test. PSA is a protein produced by cells of the prostate gland. The results fluctuate, but in general the higher they are, the more likely a man is to have prostate cancer. Gulcher's test showed that he had 2.4 nanograms of PSA per millimeter of blood, well within the normal range. Those tests are routinely recommended for men fifty years of age and older. But because Gulcher was not yet fifty, most doctors would never have given his results a second thought.

With Gulcher's genetic profile in hand, however, his physician scheduled an ultrasound, just to be sure. The films revealed an aggressive tumor, though it had not yet spread beyond the prostate. "Jeff was so young that nobody would have made anything of that kind of PSA score for ten years," Stefansson said, staring into Gulcher's empty office. "By which time he would surely have been long dead." Gulcher had surgery, quickly recovered, and returned to work. His prognosis is excellent.

Gulcher makes his living pondering the meaning of risk. Most people don't. Critics of the tests say they are still too complex for an average consumer to fully understand. Kari Stefansson disagrees. "That is such bullshit," he screeched. "We are actually criticized for revealing valuable information to unsuspecting citizens *at their request, people who paid for exactly that service.* If somebody does not want to know this information he should not have the test done. It's not required. But it is extraordinarily patronizing to tell a person that he is not mature enough to learn about himself.

"By the way," he continued, "in most American states, you can get in a car and use your driver's license as identification to buy a gun. Then you can drive to a liquor store. You can have the bottle, the gun, and the car. That's fine. But for heaven's sake don't learn anything important about yourself or your family. For some diseases there is no treatment or no useful response *yet*. But you have to remember that our ability to treat diseases was always preceded by our ability to diagnose them. So our ability to prevent diseases will surely be preceded by our ability to assess risk."

That fact is easier to handle in theory than in practice, however. Throughout the early years of the AIDS epidemic many people who had reason to fear they might be infected nevertheless didn't want to know. At the time, there was no treatment or cure. A positive test was a death sentence with no reprieve. "These decisions are never easy to make," said Arthur Caplan, the director of the Center for Bioethics at the University of Pennsylvania. "That lag between knowledge and application can be excruciating. Maybe personal genomics will look different in ten years, but right now it's a world of fortune-telling and bad news."

That depends on what you learn. If, for example, you discover that you possess a greatly increased risk of developing type 2 diabetes or heart disease, there are changes in diet and lifestyle that can help. There are also numerous medications. Will they help enough? Nobody will know until more genetic information is available. The tests have already proven their value in other ways, though. Genome-wide association tests have revealed how abnormal control of inflammation lies behind one of the principal causes of age-related macular degeneration, which is a leading cause of vision loss in Americans sixty years of age and older. More than one promising drug is already under development. The tests have also discovered genes that reveal pathways of inflammation critical for the development of inflammatory bowel disease, as well as genetic pathways for heart disease, diabetes, and obesity.

A principal goal of this research is to provide doctors with information that will take the guesswork out of writing prescriptions. In the case of the blood thinner warfarin that has already begun to happen. Warfarin is prescribed to two million people each year in the United States. The proper dose can be difficult to determine, and until recently doctors simply had to make an educated guess. Too much of the drug will put a patient at high risk for bleeding; too little can cause blood clots that lead to heart attacks. The dose can depend on age, gender, weight, and medical history. But it also depends on genetics. Two versions of the CYP2C9 gene can retard the body's ability to break down warfarin. This causes the drug's concentration in the bloodstream to decrease more slowly, which means the patient would need a lower dose. Armed with that kind of information—which these tests now provide—a physician is far more likely to get the dose right the first time.

That is the essence of pharmacogenetics. If three people out of a thousand die during a clinical trial due to a drug reaction, that drug will never make it to the market in the United States, even though it would have worked without complications for more than 99 percent of patients. If we knew who those three people were likely to be, however, none of that would matter. Obviously, that kind of knowledge would have saved thousands of lives lost to Vioxx. And it would have permitted millions who were not at risk of heart attack or stroke to continue to take a drug that had helped them immensely.

"We are just starting all this," George Church said. In addition to his academic and entrepreneurial commitments, Church advises several genomics companies, including 23andme. "But there is already great value to these tests. If you happen to have a SNP that leads to a disease that changing behavior will help, then it's magnificent. So if you have a propensity to diabetes, you're going to want to exercise, don't eat certain things, etcetera. If you have a propensity to a certain type of heart disease, etcetera, etcetera. If you have a propensity towards Alzheimer's, you might want to start on a statin early, you know?"

AFTER WALKING OUT of Church's laboratory at Harvard, I took a cab to the airport and flew home. It wasn't a pleasant flight because I couldn't stop thinking about the terrifying phrase "propensity toward Alzheimer's." Who wouldn't fear a disease that starts by making us forget much of what we would choose to remember and ends in feral despair? I have special reason to worry. A few years ago my father began to disappear into a cloud of dementia. His illness took the normal pattern—first forgetting keys (as we all do), then names, then simple directions, and eventually whatever you had told him five minutes before. Inevitably, he became incapable of fending for himself. I can think of no worse fate.

For most common diseases, the relative risks posed to individuals by specific genetic mutations remain unclear. There are just too many moving parts we have yet to analyze. Alzheimer's is an exception. Genomic studies have provided compelling evidence that a variant of at least one protein, called APOE and found on chromosome 19, dramatically increases the risk of developing the disease. APOE contains the instructions necessary to make a protein called apolipoprotein, which plays a complicated role in moderating cholesterol and clearing fats from the blood. There are three common forms, or alleles—APOE2, 3, and 4. APOE4 is the time bomb.

People with two copies of APOE4 have fifteen times the risk of developing Alzheimer's than a typical person of similar ethnic heritage. They are also at great risk of losing their memory far more rapidly than people without this allele, or those who have just one copy. The correlation between APOE4 and Alzheimer's disease is so dramatic that when James Watson became the second person (Craig Venter was the first) to publish his entire genomic sequence in 2007, he chose, of all the billions of nucleotides that comprise his DNA, to block only that data. There is Alzheimer's in Watson's family, and despite his age—he was seventy-nine at the time—Watson said he didn't wish to know the status of such a debilitating disease for which there is no cure.

Many, perhaps most, people would make the same decision, choosing to subscribe to that well-worn aphorism from Ecclesiastes: "With much wisdom comes much sorrow; the more knowledge the more grief." Others adhere to a more radical, denialist vision: "Ignorance is bliss." I prefer to see fate the way Lawrence of Arabia saw it after he managed to cross the Nefud desert. "Remember," he said to a stunned Ali, who had warned that the trip would kill Lawrence, the camels, and all his men. "Nothing is written unless you write it." It's not as if I believed that knowledge would permit me to alter my prospects of developing Alzheimer's, but it would surely permit me to alter everything else in my life.

"There's almost nothing that you can't act on in some way or another," Church had told me at Harvard. "It's probabilistic just like every decision you make in your life. What car you've got, whether to jog or not. You can always—if there's no cure, you can make a cure. You can be Augusto Odone. You can do the next *Lorenzo's Oil.* You may not be successful, but at least it will keep you busy while you're dying, or somebody in your family is dying.

"And I would definitely prefer to be busy than to be ignorant," he continued. "In other words, 'Gee, I don't know if I have the Huntington's gene, so I don't know if I should go out and raise money and get educated.' I think an increasing number of people are going to be altruistic—or selfish, depending upon how you look at it— and say, 'I want to know, so I can spend a maximum amount of time with my loved ones, fixing the family disease.'"

I had already signed up for the tests offered by Navigenics, deCODE, and 23andme. My APOE status was included on my Navigenics report, and it never occurred to me not to look.

A few days later, I poured myself a cup of coffee, sat down, and signed in to the Web site, where my data was waiting. Like the other companies, Navigenics issues a detailed guide, which it calls your Health Compass, that assesses the risks associated with many of the SNPs in your profile. (At the time it was the only company to provide customers with their APOE status, although at first it had done so by a complicated and misleading route that involved testing a different gene, one that is often inherited with APOE.)

I downloaded the 40,000-word report on my personal health. Each condition was described in three ways: as a percentile, which showed where my risks ranked compared to the sample population; as the likelihood that I would develop a given condition over my lifetime; and compared to the average person's risk. I held my breath and turned to page six, where I discovered that my lifetime risk of developing Alzheimer's—4.4 percent—was half that of the average man with my ethnic background. I don't have either APOE4 allele, which is a great relief. "You dodged a bullet," my extremely wise physician said when I told him the news. "But don't forget they might be coming out of a machine gun."

He was absolutely right. As is the case with heart disease, diabetes, autism, and many other conditions, there will almost certainly prove to be many causes of Alzheimer's. One theory holds, for example, that in some cases cholesterol may play a significant role; people with Alzheimer's often accumulate too much of a substance called amyloid precursor protein (APP). We all produce APP, but in people with Alzheimer's disease the protein gives rise to a toxic substance called beta amyloid that builds up and eventually causes plaques that kill brain cells. There remains much to be learned about this process, but some doctors recommend that people with a family history of Alzheimer's disease take statins, which help to reduce cholesterol levels even when the results from standard cholesterol tests are normal.

This is when I realized that becoming an early adopter of personal genomics isn't like buying one of the first iPods or some other cool technological gadget; there is a lot more at stake. My tests showed that I have a significantly increased risk of heart attack, diabetes, and atrial fibrillation. These are not solely diseases influenced by genetics, and effective measures exist to address at least some of those risks—diet and exercise, for instance. That's the good news. Adding that data to my family history of Alzheimer's disease suggests that it would probably make sense for me to begin taking a statin drug to lower cholesterol (even though mine is not high).

But complex as those variables are, it's still not that simple. About one person in ten thousand who take statins experiences a condition known as myopathy—muscle pain and weakness. (And since millions of Americans take the drug, those numbers are not as insignificant as they might seem.) It turns out that I have one C allele at SNP rs414056, which is located in the SLCO1B1 gene. That means I have nearly five times the chance of an adverse reaction to statins as people who have no Cs on that gene. (It could be worse; two Cs and your odds climb to seventeen times the average.) Now, what does that mean exactly? Well, if the study is correct I still have far less than a 1 percent chance of experiencing myopathy. I'll take those odds. As 23andme points out in its description of the statin response, "Please note that myopathy is a very rare side effect of statins even among those with genotypes that increase their odds of experiencing it." The risks of heart disease, however, and, in my family, Alzheimer's disease, are not rare.

CRUISING THROUGH ONE'S genomic data is not for the faint of heart. Thanks to 23andme, I now know that I am left-eyed and can taste bitter food. Cool. But I am also a slow caffeine metabolizer. That's a shame, because for people like me coffee increases the risk of heart attack, and I already have plenty of those risks.

The information, though, helps explain a mystery of my youth. I drank a lot of coffee, and periodically I would see studies that suggested coffee increased the risk of heart disease. Then other reports would quickly contradict them. With this new genetic information those differences start to make sense; some people react badly to a lot of coffee and others do not. I ended up with genetic bad luck on the caffeine front and have no choice but to drink less of it.

I'm not resistant to HIV or malaria but I am resistant, unusually enough, to the norovirus (which is the most common cause of what people think is stomach flu; actually, it's not flu at all). My maternal ancestors came from somewhere in the Urals—but I also have a bit of Berber in me because at some point seventeen thousand years ago, after the last Ice Age, my paternal line seems to have made its way into northern Africa.

If the calculations provided by these tests fail to satiate your curiosity, you can always analyze the million lines of raw data that spell your DNA (or at least as much of it as these companies currently process). You can download the data in a Zip file as if it were a song from iTunes or some family photos. Then simply plug that information into a free program called Promethease that annotates thousands of genotypes and spits back unimaginably detailed information about whatever is known about every SNP. Promethease is not for everyone, or really for very many people. It's so comprehensive that it is difficult to interpret—sort of like getting all the hits from a Google search dumped in your lap (and for most people, in a language they don't speak).

These are still early days in genomics, but it won't be long until people will carry their entire genome on their cell phone—along with an application that helps make sense of it all. When you pick up those dozen eggs at the store your phone will remind you that not only do you have high cholesterol but you have already bought eggs this week. It will warn a diabetic against a food with sugar, and a vegan to skip the soup because it was made from meat stock. It would ensure that nobody with hemochromatosis slipped up and bought spinach, and in my case, when I buy coffee beans, it would nag me to remember that they had better be decaf.

Someday—and not so long from now—medicine really will be personal. Then everyone will be a member of his own race. When that happens one has to wonder, Will discrimination finally disappear, or will it just find a new voice? That's up to us. In literature, scientific future is often heartless and grim. The 1997 film *Gattaca* was a work of science fiction about a man burdened with DNA he inherited from his parents, rather than having had it selected for him before conception. Most people were made to order. But not the main character, Vincent. He was a member of "a new genetic underclass that does not discriminate by race." A victim of genoism. As he points out in the film, "What began as a means to rid society of inheritable diseases has become a way to design your offspring—the line between health and enhancement blurred forever. Eyes can always be brighter, a voice purer, a mind sharper, a body stronger, a life longer."

Some people watched that movie and shuddered. I wasn't among them. There are many worrisome possibilities about the future, questions of privacy, equity, and personal choice not least among them. Even the most ethically complex issues can be framed positively, though, provided we are willing to discuss them. There is no reason why the past has to become the future.

"Terrible crimes have been committed in the name of eugenics. Yet I am a eugenicist," the British developmental scientist Lewis Wolpert has written. "For it now has another, very positive, side. Modern eugenics aims to both prevent and cure those with genetic disabilities. Recent advances in genetics and molecular biology offer the possibility of prenatal diagnosis and so parents can choose whether to terminate a pregnancy. There are those who abhor abortion, but that is an issue that should be kept quite separate from discussions about genetics. In Cyprus, the Greek Orthodox Church has cooperated with clinical geneticists to reduce dramatically the number of children born with the crippling blood disease thalassemia. This must be a programme that we should all applaud and support. I find it hard to think of a sensible reason why anybody should be against curing those with genetic diseases like muscular dystrophy and cystic fibrosis."

You don't have to be Dr. Frankenstein to agree with him. We need to address these issues and others we have yet to envision. There will be many ways to abuse genomics. The same technologies that save and prolong millions of lives can also be used to harm people and discriminate against them. But hasn't that always been true? The stakes are higher now, but the opportunities are greater. We are still in control of our fate, although denialists act as if we are not. The worst only happens when we let it happen.

6

Surfing the Exponential

The first time Jay Keasling remembers hearing the word "artemisinin"—about a decade ago—he had no idea what it meant. "Not a clue," Keasling, a professor of biochemical engineering at the University of California at Berkeley, recalled. Although artemisinin has become the world's most important malaria medicine, Keasling wasn't up on infectious diseases. But he happened to be in the process of creating a new discipline, synthetic biology, which, by combining elements of engineering, chemistry, computer science, and molecular biology, seeks nothing less than to assemble the biological tools necessary to redesign the living world.

No scientific achievement—not even splitting the atom—has promised so much, and none has come with greater risks or clearer possibilities for deliberate abuse. If they fulfill their promise, the tools of synthetic biology could transform microbes into tiny, self-contained factories—creating cheap drugs, clean fuels, and entirely new organisms to siphon carbon dioxide from the atmosphere we have nearly destroyed. To do that will require immense commitment and technical skill. It will also demand something more basic: as we watch the seas rise and snow-covered mountaintops melt, synthetic biology provides what may be our last chance to embrace science and reject denialism.

For nearly fifty years Americans have challenged the very idea of progress, as blind faith in scientific achievement gave way to suspicion and doubt. The benefits of new technologies—from genetically engineered food to the wonders of pharmaceuticals—have often been oversold. And denialism thrives in the space between promises and reality. We no longer have the luxury of rejecting change, however. Our only solutions lie in our skills.

Scientists have been manipulating genes for decades of course—inserting, deleting, and changing them in various molecules has become a routine function in thousands of labs. Keasling and a rapidly growing number of his colleagues have something far more radical in mind. By using gene sequence information and synthetic DNA, they are attempting to reconfigure the metabolic pathways of cells to perform entirely new functions, like manufacturing chemicals and drugs. That's just the first step; eventually, they intend to construct genes—and new forms of life—from scratch. Keasling and others are putting together a basic foundry of biological components—BioBricks, as Tom Knight, the senior research scientist from MIT who helped invent the field, has named them. Each BioBrick, made of standardized pieces of DNA, can be used interchangeably to create and modify living cells.

"When your hard drive dies you can go to the nearest computer store, buy a new one, and swap it out," Keasling said. "That's because it's a standard part in a machine. The entire electronics industry is based on a plug-and-play mentality. Get a transistor, plug it in, and off you go. What works in one cell phone or laptop should work in another. That is true for almost everything we build: when you go to Home Depot you don't think about the thread size on the bolts you buy because they're all made to the same standard. Why shouldn't we use biological parts in the same way?" Keasling and others in the field—who have formed a bicoastal cluster in the San Francisco Bay Area and in Cambridge, Massachusetts—see cells as hardware and genetic code as the software required to make them run. Synthetic biologists are convinced that with enough knowledge, they will be able to write programs to control those genetic components, which would not only let them alter nature, but guide human evolution as well.

In 2000, Keasling was looking for a chemical compound that could demonstrate the utility of these biological tools. He settled on a diverse class of organic molecules known as isoprenoids, which are responsible for the scents, flavors, and even colors in many plants: eucalyptus, ginger, and cinnamon, for example, as well as the yellow in sunflowers and red in tomatoes. "One day a graduate student stopped by and said, 'Look at this paper that just came out on amorphadiene synthase,'" Keasling told me as we sat in his in office in Emeryville, across the Bay Bridge from San Francisco. He had recently been named chief executive officer of the new Department of Energy Joint BioEnergy Institute (JBEI), a partnership between three national laboratories and three research universities, led by the Lawrence Berkeley National Laboratory. The consortium's principal goal is to design and manufacture artificial fuels that emit little or no greenhouse gases—one of President Barack Obama's most frequently cited priorities.

Keasling wasn't sure what to tell his student. "'Amorphadiene,' I said. 'What's that?' He told me that it was a precursor to artemisinin. I said, 'What's *that*?' and he said it was supposedly an effective antimalarial. I had never worked on malaria. As a microbiology student I had read about the life cycle of the falciparum parasite; it was fascinating and complicated. But that was pretty much all that I remembered. So I got to studying and quickly realized that this precursor was in the general class we were planning to investigate. And I thought, amorphadiene is as good a target as any. Let's work on that."

Malaria infects as many as five hundred million of the world's poorest people every year. For centuries the standard treatment was quinine, and then the chemically related compound chloroquine. At ten cents per treatment, chloroquine was cheap, simple to make, and it saved millions of lives. In Asia, though, by the height of the Vietnam War, the most virulent malaria parasite—falciparum—had grown resistant to the drug. Eventually, that resistance spread to Africa, where malaria commonly kills up to a million people every year, 85 percent of whom are under the age of five. Worse, the second line of treatment, sulfadoxine-pyrimethanine, or SP, had also failed widely.

Artemisinin, when taken in combination with other drugs, has become the only consistently successful treatment that remains. (Relying on any single drug increases the chances that the malaria parasite will develop resistance; if taken by itself even artemisinin poses dangers, and for that reason the treatment has already begun to fail in parts of Cambodia.) Known in the West as *Artemisia annua,* or sweet wormwood, the herb grows wild in many places, but until recently it had been used mostly in Asia. Supplies vary and so does the price, particularly since 2005, when the World Health Organization officially recommended that all countries with endemic malaria adopt artemisinin-based combination therapy as their first line of defense.

That approach, while unavoidable, has serious drawbacks: combination therapy costs ten to twenty times more than chloroquine, and despite growing assistance from international charities, that is far too much money for most Africans or their governments. In Uganda, for example, one course of artemisinin-based medicine would cost a typical family as much as it spends in two months for food. Artemisinin is not an easy crop to cultivate. Once harvested, the leaves and stems have to be processed rapidly or they will be destroyed by exposure to ultraviolet light. Yields are low, and production is expensive. Although several thousand African farmers have begun to plant the herb, the World Health Organization expects that for the next several years the annual demand—as many as five hundred million courses of treatment per year—will far exceed the supply. Should that supply disappear, the impact would be incalculable. "Losing artemisinin would set us back years—if not decades," Kent Campbell, a former chief of the malaria branch at the Centers for Disease Control, and head of the Malaria Control and Evaluation Partnership in Africa, said. "One can envision any number of theoretical public health disasters in the world. But this is not theoretical. This is real. Without artemisinin, millions of people could die."

JAY KEASLING is not a man of limited ambitions. "We have got-ten to the point in human history where we simply do not have to accept what nature has given us," he told me. It has become his motto. "We can modify nature to suit our aims. I believe that completely." It didn't take long before he realized that making amorphadiene presented an ideal way to prove his point. His goal was, in effect, to dispense with nature entirely, which would mean forgetting about artemisinin harvests and the two years it takes to turn those leaves into drugs. If each cell became its own factory, churning out the chemical required to make artemisinin, there would be no need for an elaborate and costly manufacturing pro-cess either. He wondered, why not try to build the drug out of genetic parts? How many millions of lives would be saved if, by using the tools of synthetic biology, he could construct a cell to manufacture that particular chemical, amorphadiene? It would require Keasling and his team to dismantle several different organ-isms, then use parts from nearly a dozen of their genes to cobble together a custom-built package of DNA. They would then need to create an entirely new metabolic pathway, one that did not exist in the natural world.

By 2003, the team reported its first success, publishing a pa-per in *Nature Biotechnology* that described how they constructed that pathway—a chemical circuit the cell needs to do its job—by inserting genes from three organisms into *E. coli*, one of the world's most common bacteria. The paper was well received, but it was only the first step in a difficult process; still, the research helped Keasling secure a $42.6 million grant from the Bill and

Melinda Gates Foundation. It takes years, millions of dollars, much effort, and usually a healthy dose of luck to transform even the most ingenious idea into a product you can place on the shelf of your medicine cabinet. Keasling wasn't interested in simply proving the science worked; he wanted to do it on a scale that would help the world fight malaria. "Making a few micrograms of artemisinin would have been a neat scientific trick," he said. "But it doesn't do anybody in Africa any good if all we can do is a cool experiment in a Berkeley lab. We needed to make it on an industrial scale."

To translate the science into a product, Keasling helped start a company, Amyris Biotechnologies, to refine the raw organism, then figure out how to produce it more efficiently. Slowly, the company's scientists coaxed greater yields from each cell. What began as 100 micrograms per liter of yeast eventually became 25 grams per liter. The goal was to bring the cost of artemisinin down from more than ten dollars a course to less than one dollar. Within a decade, by honing the chemical sequences until they produced the right compound in the right concentration, the company increased the amount of artemisinic acid that each cell could produce by a factor of one million. Keasling, who makes the cellular toolkit available to other researchers at no cost, insists that nobody profit from its sale. (He and the University of California have patented the process in order to make it freely available.) "I'm fine with earning money from research in this field," he said. "I just don't think we need to profit from the poorest people on earth."

Amyris then joined the nonprofit Institute for OneWorld Health, in San Francisco, and in 2008 they signed an agreement with the Paris-based pharmaceutical company Sanofi-Aventis to produce the drug, which they hope to have on the market by the end of 2011. Scientific response has been largely reverential—it is, after all, the first bona fide product of synthetic biology, proof of a principle that we need not rely on the unpredictable whims of nature to address the world's most pressing crises. But there are those who wonder what synthetic artemisinin will mean for the thousands of farmers who have begun to plant the crop. "What happens to struggling farmers when laboratory vats in California replace [wormwood] farms in Asia and East Africa?" asked Jim Thomas, an activist with ETC Group, a technology watchdog based in Canada. Thomas has argued that while the science of synthetic biology has advanced rapidly, there has been little discussion of the ethical and cultural implications involved in altering nature so fundamentally, and he is right. "Scientists are making strands of DNA that have never existed," Thomas said. "So there is nothing to compare them to. There's no agreed mechanisms for safety, no policies."

Keasling, too, believes we need to have a national conversation about the potential impact of this technology, but he is mystified by opposition to what would be the world's most reliable source of cheap artemisinin. "We can't let what happened with genetically engineered foods"—which have been opposed by millions of people for decades—"happen again," he said. "Just for a moment imagine that we replaced artemisinin with a cancer drug. And let's have the entire Western world rely on some farmers in China and Africa who may or may not plant their crop. And let's have a lot of American children die because of that. It's so easy to say, 'Gee, let's take it slow' about something that can save a child thousands of miles away. I don't buy it. They should have access to Western technology just as we do. Look at the world and tell me we shouldn't be doing this. It's not people in Africa who see malaria who say, 'Whoa, let's put the brakes on.'"

Keasling sees artemisinin as the first part of a much larger program. "We ought to be able to make any compound produced by a plant inside a microbe," he said. "We ought to have all these metabolic pathways. You need this drug? okay, we pull this piece, this part, and this one off the shelf. You put them into a microbe and two weeks later out comes your product."

That's the approach Amyris has taken in its efforts to develop new fuels. "Artemisinin is a hydrocarbon and we built a microbial platform to produce it," Keasling said. "We can remove a few of the genes to take out artemisinin and put in a different hydrocarbon to make biofuels." Amyris, led by John Melo, who spent years as a senior executive at British Petroleum, has already engineered three molecules that can convert sugar to fuel. "It is thrilling to address problems that only a decade ago seemed insoluble," Keasling said. "We still have lots to learn and lots of problems to solve. I am well aware that makes people anxious, and I understand why. Anything so powerful and new is troubling. But I don't think the answer to the future is to race into the past."

FOR THE FIRST four billion years, life on earth was shaped entirely by nature. Propelled by the forces of selection and chance, the most efficient genes survived and evolution ensured they would thrive. The long, beautiful Darwinian process of creeping forward by trial and error, struggle and survival, persisted for millennia. Then, about ten thousand years ago, our ancestors began to gather in villages, grow crops, and domesticate animals. That led to new technologies—stone axes and looms, which in turn led to better crops and the kind of varied food supply that could support a larger civilization. Breeding goats and pigs gave way to the fabrication of metal and machines. Throughout it all, new species, built on the power of their collected traits, emerged, while others were cast aside.

As the world became larger and more complex, the focus of our discoveries kept shrinking—from the size of the planet, to a species, and then to individual civilizations. By the beginning of the twenty-first century we had essentially become a society fixated on cells. Our ability to modify the smallest components of life through molecular biology has endowed humans with a power that even those who exercise it most proficiently cannot claim to fully comprehend. Man's mastery over nature has been predicted for centuries—Bacon insisted on it, Blake feared it profoundly. Little more than one hundred years have passed, however, since Gregor Mendel demonstrated that the defining characteristics of a pea plant—its shape, size, and the color of the seeds, for example—are transmitted from one generation to the next in ways that can be predicted, repeated, and codified.

Since then, the central project of biology has been to break that code and learn to read it—to understand how DNA creates and perpetuates life. As an idea, synthetic biology has been around for many years. It took most of the past century to acquire the knowledge, develop the computing power, and figure out how to apply it all to DNA. But the potential impact has long been evident. The physiologist Jacques Loeb was perhaps the first to predict that we would eventually control our own evolution by creating and manipulating new forms of life. He considered artificial synthesis of life the "goal of biology," and encouraged his students to meet that goal. In 1912, Loeb, one of the founders of modern biochemistry, wrote that "nothing indicates . . . that the artificial production of living matter is beyond the possibilities of science. . . . We must succeed in producing living matter artificially or we must find the reasons why this is impossible."

The Nobel Prize–winning geneticist Hermann J. Muller attempted to do that. By demonstrating that exposure to X-rays can cause mutations in the genes and chromosomes of living cells, he was the first to prove that heredity could be affected by something other than natural selection. He wasn't entirely certain that humanity would use the information responsibly, though. "If we did attain to any such knowledge or powers there is no doubt in my mind that we would eventually use them," Muller wrote in 1916. "Man is a megalomaniac among animals—if he sees mountains he will try to imitate them by building pyramids, and if he sees some grand process like evolution, and thinks it would be at all possible for him to be in on that game, he would irreverently have to have his whack at that too."

We have been having that "whack" ever since. Without Darwin's most important—and contentious—contribution, none of it would have been possible, because the theory of evolution explained that every species on earth is related in some way to every other species; more important, we carry a record of that history in each of our bodies. In 1953, James Watson and Francis Crick began to make it possible to understand why, by explaining how DNA arranges itself. The language of just four chemical letters— adenine, guanine, cytosine, and thymine—comes in the form of enormous chains of nucleotides. When joined together, the arrangement of their sequences determine how each human differs from each other and from all other living beings.

By the 1970s, recombinant DNA technology permitted scientists to cut long, unwieldy molecules of nucleotides into digestible sentences of genetic letters and paste them into other cells. Researchers could suddenly combine the genes of two creatures that would never have been able to mate in nature. In 1975, concerned about the risks of this new technology, scientists from around the world convened a conference in Asilomar, California. They focused primarily on laboratory and environmental safety, and concluded that the field required only minimal regulation. (There was no real discussion of deliberate abuse—at the time it didn't seem necessary.)

In retrospect at least, Asilomar came to be seen as an intellectual Woodstock, an epochal event in the history of molecular biology. Looking back nearly thirty years later, one of the conference's organizers, the Nobel laureate Paul Berg, wrote that "this unique conference marked the beginning of an exceptional era for science and for the public discussion of science policy. Its success permitted the then contentious technology of recombinant DNA to emerge and flourish. Now the use of the recombinant DNA technology dominates research in biology. It has altered both the way questions are formulated and the way solutions are sought."

Scientists at the meeting understood what was at stake. "We

can outdo evolution," said David Baltimore, genuinely awed by this new power to explore the vocabulary of life. Another researcher joked about joining duck DNA with orange DNA. "In early 1975, however, the new techniques hardly aspired to either duck or orange DNA," Michael Rogers wrote in the 1977 book *Biohazard*, his riveting account of the meeting at Asilomar and of the scientists' attempts to confront the ethical as well as biological impact of their new technology. "They worked essentially only with bacteria and viruses—organisms so small that most human beings only noticed them when they make us ill."

That was precisely the problem. Promising as these techniques were, they also made it possible for scientists to transfer viruses—and cancer cells—from one organism to another. That could create diseases anticipated by no one and for which there would be no natural protection, treatment, or cure. The initial fear "was not that someone might do so on purpose," Rogers wrote—that would come much later—"but rather that novel microorganisms would be created and released altogether accidentally, in the innocent course of legitimate research."

Decoding sequences of DNA was tedious work. It could take a scientist a year to complete a stretch ten or twelve base pairs long (our DNA consists of three billion such pairs). By the late 1980s automated sequencing had simplified the procedure, and today machines are capable of processing that information, and more, in seconds. Another new tool—polymerase chain reaction—was required to complete the merger of the digital and biological worlds. Using PCR, a scientist can take a single DNA molecule and copy it many times, making it easier to read and manipulate.

That permits scientists to treat living cells like complex packages of digital information that happen to be arranged in the most elegant possible way.

Mixing sequences of DNA, even making transgenic organisms, no longer requires unique skills. The science is straightforward. What came next was not. Using the tools of genomics, evolutionary biology, and virology, researchers began to bring dead viruses back to life. In France, the biologist Thierry Heidmann took a virus that had been extinct for hundreds of thousands of years, figured out how the broken parts were originally aligned, and then pieced them back together. After resurrecting the virus, which he named Phoenix, he and his team placed it in human cells and found that their creation could insert itself into the DNA of those cells. They also mixed the virus with cells taken from hamsters and cats. It quickly infected them all, offering the first evidence that the broken parts of an ancient virus could once again be made infectious.

As if experiments like those were not sufficient to conjure images of Frankenstein's monster or *Jurassic Park*, researchers have now resurrected the DNA of the Tasmanian tiger, the world's largest carnivorous marsupial, which has been extinct for more than seventy years. In 2008, scientists from the University of Melbourne in Australia and the University of Texas M. D. Anderson Cancer Center in Houston extracted DNA from two strands of tiger hair that had been preserved in museums. They inserted a fragment of a tiger's DNA that controlled the production of collagen into a mouse embryo. That switched on just the right gene, and the embryo began to churn out collagen—marking the first time that material from an extinct creature (other than a virus) has functioned inside a living cell.

It will not be the last. A team from Pennsylvania State University, working with fossilized hair samples from a 65,000-year-old woolly mammoth, has already figured out how to modify that DNA and place it inside an elephant's egg. The mammoth could then be brought to term in an elephant mother. "There is little doubt that it would be fun to see a living, breathing woolly mammoth—a shaggy, elephantine creature with long curved tusks who reminds us more of a very large, cuddly stuffed animal than of a T. rex," the *New York Times* wrote in an editorial after the discovery was announced. "We're just not sure that it would be all that much fun for the mammoth." The next likely candidates for resurrection are our ancient relatives, the Neanderthals, who were probably driven to extinction by the spread of modern humans into Europe some forty thousand years ago.

All of that has been a prelude—technical tricks from a youthful discipline. The real challenge is to create a synthetic organism made solely from chemical parts and blueprints of DNA. In the early 1990s, working at his nonprofit organization the Institute for Genomic Research, Craig Venter and his colleague Clyde Hutchison began to wonder whether they could pare life to its most basic components and then try to use those genes to create a synthetic organism they could program. They began modifying the genome of a tiny bacterium called *Mycoplasma genitalium*, which contained 482 genes (humans have about 23,000) and 580,000 letters of genetic code, arranged on one circular chromosome—the smallest genome of any known natural organism. Venter and his colleagues then systematically removed genes, one by one, to find the smallest set that could sustain life.

He called the experiment the Minimal Genome Project. By the

beginning of 2008, Venter's team had pieced together thousands of chemically synthesized fragments of DNA and assembled a new version of the organism. Then, using nothing but chemicals, they produced the entire genome of *M. genitalium* from scratch. "Nothing in our methodology restricts its use to chemically synthesized DNA," Venter noted in the report of his work, which was published in *Science* magazine. "It should be possible to assemble any combination of synthetic and natural DNA segments in any desired order." That may turn out to be one of the most memorable asides in the history of science. Next, he intends to transplant the artificial chromosome into the walls of another cell, and then "boot it up," to use his words—a new form of life that would then be able to replicate its own DNA, the first truly artificial organism. Venter has already named the creation Synthia. He hopes that Synthia, and similar products, will serve essentially as vessels that can be modified to carry different packages of genes. One package might produce a specific drug, for example, and another could have genes programmed to digest excess carbon in the atmosphere.

In 2007, the theoretical physicist and intellectual adventurer Freeman Dyson took his grandchildren to the Philadelphia Flower Show and then the Reptile Super Show in San Diego. "Every orchid or rose or lizard or snake is the work of a dedicated and skilled breeder," he wrote in an essay for the *New York Review of Books*. "There are thousands of people, amateurs and professionals, who devote their lives to this business." This, of course, we have been doing in one way or another for millennia. "Now imagine what will happen when the tools of genetic engineering become accessible to these people."

He didn't say if, he said when: because it is only a matter of time until domesticated biotechnology presents us with what Dyson describes as an "explosion of diversity of new living creatures. . . . Designing genomes will be a personal thing, a new art form as creative as painting or sculpture. Few of the new creations will be masterpieces, but a great many will bring joy to their creators and variety to our fauna and flora."

Biotech games, played by children "down to kindergarten age but played with real eggs and seeds," could produce entirely new species, as a lark. "These games will be messy and possibly dangerous," he wrote. "Rules and regulations will be needed to make sure that our kids do not endanger themselves and others. The dangers of biotechnology are real and serious."

I have never met anyone engaged in synthetic biology who would disagree. Venter in particular has always stressed the field's ethical and physical risks. His current company, Synthetic Genomics, commissioned a lengthy review of the ethical implications of the research more than a year before the team even entered the lab. How long will it be before proteins engineer their own evolution? "That's hard to say," Venter told me, "but in twenty years this will be second nature for kids. It will be like Game Boy or Internet chat. A five-year-old will be able to do it."

Life on earth proceeds in an arc—one that began with the Big Bang, and evolved to the point where a smart teenager is capable of inserting a gene from a cold-water fish into a strawberry to help protect it from the frost. You don't have to be a Luddite or Prince Charles—who famously has foreseen a world reduced to "grey goo" by avaricious and out-of-control technology—to recognize that synthetic biology, if it truly succeeds, will make it possible to supplant the world created by Darwinian evolution with a world created by us.

"Many a technology has at some time or another been deemed an affront to God, but perhaps none invites the accusation as directly as synthetic biology," the editors of *Nature*—who none-theless support the technology—wrote in 2007. "Only a deity predisposed to cut-and-paste would suffer any serious challenge from genetic engineering as it has been practiced in the past. But the efforts to design living organisms from scratch—either with a wholly artificial genome made by DNA synthesis technology or, more ambitiously, by using non-natural, bespoke molecular machinery—really might seem to justify the suggestion" that "for the first time, God has competition."

"WHAT IF WE could liberate ourselves from the tyranny of evolu-tion by being able to design our own offspring?" Drew Endy asked the first time we met. It was a startling question—and it was meant to startle. Endy is synthetic biology's most compelling evangelist. He is also perhaps its most disturbing, because, while he displays a childlike eagerness to start building new creatures, he insists on discussing both the prospects and dangers of this new science in nearly any forum he can find. "I am talking about building the stuff that runs most of the living world," he said. "If this is not a national strategic priority, what possibly could be?"

Endy, who was trained as a structural engineer, is almost always talking about designing or building something. He spent his youth fabricating worlds out of Lincoln Logs and Legos. What he would like to build now are living organisms. We were sitting in his office at the Massachusetts Institute of Technology, where until the spring of 2008, he was assistant professor in the recently formed department of biological engineering. (That summer, he moved to Stanford.) Perhaps it was the three well-worn congas sitting in the corner of Endy's office, the choppy haircut that looked like something he might have gotten in a treehouse, or the bicycle dangling from his wall, but when he speaks about putting new forms of life together, it's hard not to think of that boy and his Legos.

I asked Endy to describe the implications of the field and why he thought so many people are repelled by the idea of creating new organisms. "Because it's scary as hell," he said. "It's the coolest platform science has ever produced—but the questions it raises are the hardest to answer. For instance, now that we can sequence DNA, what does that mean?" Endy argues that if you can sequence something properly and you possess the information for describing that organism—whether it's a virus, a dinosaur, or a human—you will eventually be able to construct an artificial version of it.

"That gives us an alternate path for propagating living organisms," he said. "The natural path is direct descent from a parent—from one generation to the next. But that is an error process—there are mistakes in the code, many mutations," although in Darwin's world a certain number of those mutations are necessary. "If you could complement evolution with a secondary path—let's decode a genome, take it offline to the level of information"; in other words, let's break it down to its specific sequences of DNA the way we would the code in a software program—"we can then design whatever we want, and recompile it. At that point you can make disposable biological systems that don't have to produce offspring, and you can make much simpler organisms."

Endy stopped long enough for me to digest the fact that he was talking about building our own children, not to mention alternate versions of ourselves. Humans are almost unimaginably complex, but if we can bring a woolly mammoth back to life or create and "boot up" a synthetic creature made from hundreds of genes, it no longer seems impossible, or even improbable, that scientists will eventually develop the skills to do the same thing with our species. "If you look at human beings as we are today, one would have to ask how much of our own design is constrained by the fact that we have to be able to reproduce," Endy said. "In fact, those constraints are quite significant. But by being able to design our own offspring we can free ourselves from them. Before we talk about that, however, we have to ask two critical questions: what sorts of risks does that bring into play, and what sorts of opportunities?"

The deeply unpleasant risks associated with synthetic biology are not hard to contemplate: who would control this technology, who would pay for it, and how much would it cost? Would we all have access or, as in *Gattaca*, would there be genetic haves and have-nots? Moreover, how safe can it be to manipulate and create life? How likely are accidents that would unleash organisms onto a world that is not prepared for them? And will it be an easy technology for people bent on destruction to acquire? After all, if Dyson is right and kids will one day design cute backyard dinosaurs, it wouldn't take much imagination for more malevolent designers to create organisms with radically different characteristics. "These are things that have never been done before," said Endy. "If the society that powered this technology collapses in some way we would go extinct pretty quickly. You wouldn't have a chance to revert back to the farm or to the prefarm. We would just be gone."

Those fears have existed since we began to transplant genes in crops. They are the principal reason why opponents of genetically engineered food invoke the precautionary principle, which argues that potential risks must always be given more weight than possible benefits. That is certainly the approach suggested by people like Thomas of ETC, who describes Endy as "the alpha Synthusiast." But he added that Endy was also a reflective scientist who doesn't discount the possible risks of his field. "To his credit, I think he's the one who's most engaged with these issues," Thomas said. Endy hopes that's true, but doesn't want to relive the battles over genetically engineered food, where the debate has so often focused on theoretical harm rather than tangible benefits. "If you build a bridge and it falls down you are not going to be permitted to design bridges ever again," he said. "But that doesn't mean we should never build a new bridge. There, we have accepted the fact that risks are inevitable. When it comes to engineering biology, though, scientists have never developed that kind of contract with society. We obviously need to do that."

Endy speaks with passion about the biological future; but he also knows what he doesn't know. And what nobody else knows either. "It is important to unpack some of the hype and expectation around what you can do with biotechnology as a manufacturing platform," he said. "We have not scratched the surface—but how far will we be able to go? That question needs to be discussed openly. Because you can't address issues of risk and society unless you have an answer. If we do not frame the discussion properly we will soon face a situation where people say: Look at these scientists doing all these interesting things that have only a limited impact on our civilization, because the physics don't scale. If that is the case, we will have a hard time convincing anybody we ought to be investing our time and money this way."

The inventor and materials scientist Saul Griffith has estimated that between fifteen and eighteen terawatts of energy are required to power our planet. How much of that could we manufacture with the tools of synthetic biology? "The estimates run between five and ninety terawatts," Endy said. "And you can figure out the significance of that right away. If it turns out to be the lower figure we are screwed. Because why would we take these risks if we cannot create much energy? But if it's the top figure then we are talking about producing five times the energy we need on this planet and doing it in an environmentally benign way. The benefits in relation to the risks of using this new technology would be unquestioned. But I don't know what the number will be and I don't think anybody *can* know at this point. At a minimum then, we ought to acknowledge that we are in the process of figuring that out and the answers won't be easy to provide.

"It's very hard for me to have a conversation about these issues," he continued. "Because people adopt incredibly defensive postures. The scientists on one side and civil society organizations on the other. And to be fair to those groups, science has often proceeded by skipping the dialogue. But some environmental groups will say, 'Let's not permit any of this work to get out of a laboratory until we are sure it is all safe.' And as a practical matter that is not the way science works. We can't come back decades later with an answer. We need to develop solutions by doing them. The potential is great enough, I believe, to convince people it's worth the risk.

"We also have to think about what our society needs and what this science might do," he continued. "We have seen an example with artemisinin and malaria. That's only a first step—maybe we could avoid diseases completely. That could require us to go through a transition in medicine akin to what happened in environmental science and engineering after the end of World War II. We had industrial problems and people said, 'Hey, the river's on fire—let's put it out.' And after the nth time of doing that people started to say, 'Maybe we shouldn't make factories that put shit into the river. So let's collect all the waste.' That turns out to be really expensive because then we have to dispose of it. Finally, people said, 'Let's redesign the factories so that they don't make that crap.'" (In fact, the fire that erupted just outside Cleveland, Ohio, on the Cuyahoga River in June 1969 became a permanent symbol of environmental disaster. It also helped begin a national discussion that ended in the passage of the Clean Water Act, the Safe Drinking Water Act, and many other measures.)

"Let's say I was a whimsical futurist," said Endy—although there is nothing whimsical about his approach to science or to the future. "We are spending trillions of dollars on health care. Preventing disease is obviously more desirable than treating it. My guess is that our ultimate solution to the crisis of health care costs will be to redesign ourselves so that we don't have so many problems to deal with. But note," he stressed, "you can't possibly begin to do something like this if you don't have a value system in place that allows you to map concepts of ethics, beauty, and aesthetics onto our own existence.

"We need to understand the ways in which those things matter to us because these are powerful choices. Think about what happens when you really can print the genome of your offspring. You could start with your own sequence, of course, and mash it up with your partner, or as many partners as you like. Because computers won't care. And if you wanted evolution you can include random number generators" which would have the effect of introducing the element of chance into synthetic design.

I wondered how much of this was science fiction, and how much was genuinely likely to happen. Endy stood up. "Can I show you something?" he asked as he walked over to a bookshelf and grabbed four gray bottles. Each contained about half a cup of sugar and had a letter on it: A, T, C, or G, for the four nucleotides in our DNA. "You can buy jars of these chemicals that are derived from sugarcane," he said. "And they end up being the four bases of DNA in a form that can be readily assembled. You hook the bottles up to a machine, and into the machine comes information from a computer, a sequence of DNA—like TAATAGCAA. You program in whatever you want to build and that machine will stitch the genetic material together from scratch. This is the recipe: you take information and the raw chemicals and compile genetic material. Just sit down at your laptop and type the letters and out comes your organism."

He went to a whiteboard in the office. "This is a little bit of math I did a number of years ago which freaked me out," he said, and then went on to multiply the molecular weight of the nucleotides in those four bottles by the number of base pairs in a human genome. He used ten billion people instead of the current world population of six and a half billion and also rounded up the number of bases to ten billion, just to be conservative. "What we come out with in these bottles," he said, placing one on the palm of his left hand, "is sixty times more material than is necessary to reconstruct a copy of every human's genome on this planet."

Endy shrugged. Of course, we don't have machines that can turn those sugars into genetic parts yet—"but I don't see any physical reason why we won't," he said. "It's a question of money. If somebody wants to pay for it then it will get done." He looked at his watch, apologized, and said, "I'm sorry we will have to continue this discussion another day because I have an appointment with some people from the Department of Homeland Security."

I was a little surprised. Why? I asked.

"They are asking the same questions as you," he said. "They want to know how far is this really going to go."

The next time I saw Endy, a few months later, was in his new office at Stanford. The Bay Area is rapidly becoming as central to synthetic biology as it has always been to the computer industry. Endy looked rattled. "I just drove across the Golden Gate Bridge," he said. "The whole way I kept thinking, 'How does this thing stay up?'" I was confused. Drew Endy is a structural engineer. If crossing a bridge—particularly that bridge—worries him, who wouldn't it worry? "Look," he said, "there is uncertainty in the world. When it comes to engineering biology we don't, for the most part, know enough to do useful things. But we also don't know enough about physics. Gravity has not worked out perfectly—and yet we have the Golden Gate Bridge. And that's okay because I am comfortable with learning by doing. Biology is a really impressive manufacturing platform. And we suck at it. But if we wait until we accumulate all the knowledge, we will accomplish nothing."

Endy made his first mark on the world of biology by nearly failing the course in high school. "I got a D," he said. "And I was lucky to get it." While pursuing his engineering degree at Lehigh University, not far from Valley Forge, Pennsylvania, where he was raised, Endy took a course in molecular genetics. Because he saw the world through the eyes of an engineer he looked at the parts of cells and decided it would be interesting to try and build one. He spent his years in graduate school modeling bacterial viruses, but they are complex and Endy craved simplicity. That's when he began to think about putting cellular components together.

In 2005, never forgetting the secret of Legos—they work because you can take any single part and attach it to any other— Endy and colleagues on both coasts started the BioBricks Foundation, a nonprofit organization formed to register and develop standard parts for assembling DNA. (Endy is not the only biologist, nor even the only synthetic biologist, to translate a youth spent with blocks into a useful scientific vocabulary. "The notion of pieces fitting together—whether those pieces are integrated circuits, microfluidic components, or molecules—guides much of what I do in the laboratory," the physicist and synthetic biologist Rob Carlson wrote in his 2009 book *Biology Is Technology: The Promise, Peril, and Business of Engineering Life*, "and some of my best work has come together in my mind's eye accompanied by what I swear was an audible click.")

BioBricks, then, have become the thinking man's Lego system. The registry is a physical repository, but also an online catalog.

If you want to construct an organism, or engineer it in new ways, you can go to the site in much the same way you would buy lumber or industrial pipes online. The constituent parts of DNA—promoters, ribosomes, plasmid backbones, and thousands of other components—are cataloged, explained, and discussed. It is a kind of Wikipedia of future life forms—with the added benefit of actually providing the parts necessary to build them.

Endy argues that scientists skipped a step, at the birth of biotechnology thirty-five years ago, moving immediately to products without first focusing on the tools necessary to make them. Using standard biological parts, a synthetic biologist or biological engineer can already to some extent program living organisms in the same way a computer scientist can program a computer. The analogy doesn't work perfectly, though, because genetic code is not as linear as computer code. Genes work together in ways that are staggeringly complex and therefore difficult to predict; proteins produced by one gene will counteract or enhance those made by another. We are far from the point where we can yank a few genes off the shelf, mix them together, and produce a variety of products. But the registry is growing rapidly—and so is the knowledge needed to drive the field forward.

Research in Endy's lab has largely been animated by his fascination with switches that turn genes on and off. He and his students are attempting to create genetically encoded memory systems. His current goal is to construct a cell that can count to about 256—a number based on basic computer code. Solving the practical challenges will not be easy, since cells that count will need to send reliable signals when they divide and remember that they did.

"If the cells in our bodies had a little memory, think what we could do," Endy said. I wasn't quite sure what he meant. "You have memory in your phone," he explained. "Think of all the information it allows you to store. The phone and the technology on which it is based do not function inside cells. But if we could count to two hundred, using a system that was based on proteins and DNA and RNA, well now, all of a sudden we would have a tool that gives us access to computing and memory that we just don't have.

"Do you know how we study aging?" he continued. "The tools we use today are almost akin to cutting a tree in half and counting the rings. But if the cells had a memory we could count properly. Every time a cell divides just move the counter by one. Maybe that will let me see them changing with a precision nobody can have today. Then I could give people controllers to start retooling those cells. Or we could say, 'Wow, this cell has divided two hundred times, it's obviously lost control of itself and become cancer. Kill it.' That lets us think about new therapies for all kinds of diseases."

Synthetic biology is changing so rapidly that predictions seem pointless. Even that fact presents people like Endy with a new kind of problem. "Wayne Gretzky once famously said, 'I skate to where the puck is going, not to where the puck is.' That's what you do to become a great hockey player," Endy said. "But where do you skate when the puck is accelerating at something that seems like the speed of light, when the trajectory is impossible to follow? Who do you hire and what do we ask them to do? Because what preoccupies our finest minds today will be a seventh-grade science project in five years. Or three years.

"That is where we are with this technology. The thrill is real—but so are the fears. We are surfing an exponential now, and even for people who pay attention, surfing an exponential is a really tricky thing to do. And when the exponential you are surfing has the capacity to impact the world in such a fundamental way, in ways we have never before considered, what do you do then? How do you even talk about that?"

IN AUGUST 2002, *Science* magazine published a report titled "Chemical Synthesis of Poliovirus cDNA." It began with an assertion few virologists would dispute: "Research on viruses is driven not only by an urgent need to understand, prevent, and cure viral disease. It is also fueled by a strong curiosity about the minute particles that we can view both as chemicals and as 'living' entities." That curiosity led a team directed by Eckard Wimmer at Stony Brook University to stitch together hundreds of DNA fragments, most of which were purchased on the Internet, and then use them to build a fully functioning polio virus. The scientists then injected the virus into mice, which promptly became paralyzed and died.

The experiment, the first in which a virus was created in a laboratory solely from chemicals, caused outrage "This is a blueprint that could conceivably enable terrorists to inexpensively create human pathogens," Representative Dave Weldon said at the time; he and five other members of Congress introduced a resolution criticizing the American Association for the Advancement of Science, which publishes *Science* magazine, for publishing the study. Many scientists considered the research an irresponsible stunt. Then, in 2005, federal scientists deciphered the genetic code of the 1918 flu virus, which killed at least fifty million people, and reconstructed that virus too.

A renowned virologist once described Wimmer's polio research to me as nothing more than "proof of principle for bioterrorism," a comment I used in an article about scientists who were bringing ancient viruses back to life. He said the report would serve only to remind people how easily they could obtain the various components required to make a virus. After all, anyone can order strands of DNA over the Internet from scores of companies, nearly all of which will deliver via Federal Express. Soon after the article was published, I received a polite e-mail from Wimmer, who said I had completely misunderstood the purpose of his work, and he invited me to his laboratory to discuss it.

Wimmer met me at the door to his office, a thin, elegant man in a maroon turtleneck, gray flannel pants, and a blue cashmere sweater. He had chalked out various viral particles on the whiteboard. "I want to say before anything else that we didn't do this work to show we were good at chemistry," he told me. "First— and I think this is important—people need to know what is possible. It's not as if any smart kid out there can make polio or smallpox in their homes. These are complicated viruses—yet there seems to be this idea floating around that you can just order DNA and whip up a virus as if it were a cake. That is untrue. Could somebody who wants to hurt people make such a virus? Of course. Will one be made? I don't know, but silence isn't going to help us prevent it or respond. We need to be talking."

It didn't take long for me to realize that he was right. Synthetic biology will never fulfill its promise unless it is discussed and understood by the society it is designed to serve. If not, the cycle of opposition and denialism will begin anew. Scientists will insist that research is safe and the benefits clear. A chorus will respond: how do you know? Wimmer was right about the difficulty of making viruses too, particularly in the quantities necessary for a weapon. He wouldn't put it this way, but he believes the best defense is an offense. To protect ourselves from new diseases, including those introduced purposefully, we will need vaccines that can stop them. And to do that, scientists must understand how the parts work. (Which in the end has been the goal of his polio research.)

Before meeting Wimmer I had asked Drew Endy what he thought of the controversial research. Endy, whose fundamental approach to biological engineering is to learn by doing, has also tried to synthesize novel viruses to better understand how they work. "If it was just a single virus then I could see people wondering why he did it," Endy said. "But if you look at the arc of Eckard's research he has used synthesis to make viruses that have hundreds of mutations which attenuate their activity. Those changes can help lead to rapid vaccine responses." Vaccines are made in a couple of basic ways. Live, attenuated vaccines are often the most effective; they are composed of a virus that has been weakened or altered in order to reduce its ability to cause disease, but they can take years to develop. Wimmer introduced a modern version of that approach: a synthesized virus that had been mutated could train antibodies without causing harm. Indeed, the Defense Advanced Research Projects Agency (DARPA) has a program under way to develop vaccines "on demand," in large quantity, and at low cost, to interdict both established and new biological threats.

"You have to remember," Wimmer said in reference to his original paper, "2002 was a super-scary time after 9/11 and the anthrax attacks. I think the fear that people expressed was in not knowing the goals of the research. By 2005, people seemed more comfortable with the idea that there was a legitimate reason to reconstruct something like the 1918 flu virus in order to create a vaccine. With polio, which really doesn't affect people, it is still harder to explain that we use the research to make vaccines.

"But our approach was to remodel the virus," he went on. "I have said before—and this is true of synthetic biology in general— we have to understand that it provides wonderful solutions to terrible problems. And it can also lead to the synthesis of smallpox and polio." Many of Wimmer's original critics have come around to his point of view. In 2008, he was elected a fellow of the American Association for the Advancement of Science for "discovering the chemical structure of the poliovirus genome, elucidating genetic functions in poliovirus replication and pathogenesis, and synthesizing poliovirus de novo."

Wimmer's polio research did spark a discussion about whether synthetic biology could be used for bioterrorism; the answer, of course, is yes. If a group of well-trained scientists want to manufacture polio—or even the more complicated smallpox virus—they will be able to do so. (It should be noted, but often is not, that an evil scientist—or country—does not need fancy new technology or much money to cause widespread terror and death. Anthrax spores exist naturally in the soil. They can be extracted, grown, and turned into remarkably effective weapons with far less effort than it would take to create a lethal organism from scratch.) While creating deadly viruses from modern tools—or using them to revive smallpox—presents a compelling horror story (and rightfully so), more prosaic weapons, both biological and conventional, are easier to use, highly effective, and more accessible. "It doesn't take the fanciest technology to cause destruction," Wimmer said. "I think we all saw that on September 11."

FOR DECADES, people have described the exponential growth of the computer industry by invoking Moore's law. In 1965, Gordon Moore predicted the number of transistors that could fit onto a silicon chip would double every eighteen months, and so would the power of computers. When the IBM 360 computer was released in 1964, the top model came with eight megabytes of main memory, and it took enough space to fill a room. With a full complement of expensive components the computer could cost more than $2 million. Today, cell phones with a thousand times the memory can be purchased for less than a hundred dollars.

In 2001, Rob Carlson, then a researcher at the University of Washington and one of synthetic biology's most consistently provocative voices, decided to examine a similar phenomenon: the speed at which the capacity to synthesize DNA was growing. What he produced has come to be known as the Carlson Curve, which mirrors Moore's law, and has even begun to exceed it. Again, the effect has been stunning. Automated gene synthesizers that cost $100,000 a decade ago now cost less than $10,000. Most days, at least a dozen used synthesizers are for sale on eBay—for less than $1,000.

As the price of processing DNA drops, access (and excitement) rises. Between 1977, when Frederick Sanger published the first paper on automatic gene sequencing, and 1995, when Craig Venter published the first bacterial genome sequence, the field moved slowly. It took the next six years to complete the first draft of the immeasurably more complex human genome, and six years after that, in 2007, scientists on three continents began mapping the full genomes of one thousand people. George Church's Personal Genome Project now plans to sequence one hundred thousand. (Church is convinced that, in exchange for advertising, companies will soon make genomes available to anyone for free—a model that has certainly worked for Google.) His lab has been able to sequence billions of DNA base pairs in the time it would have taken Sanger to sequence one. "This is not because George or Craig Venter got ten billion times smarter in fifteen years," Endy said. "It's because the capacity of the tools have exploded."

In 2004, when he was still at MIT, Endy and his colleagues Tom Knight and Randy Rettberg founded iGEM, the International Genetically Engineered Machine competition, whose purpose is to promote the building of biological systems from standard parts like those in the BioBricks registry. In 2006, a team of Endy's undergraduate students used those tools to genetically reprogram *E. coli* (which normally smells awful) to smell like wintergreen while it grows and like bananas when it is finished growing. They named their project Eau d'E Coli. By 2008, with hundreds of students from dozens of countries participating, the winning team—a group from Slovenia—used biological parts that they had designed to create a vaccine for the stomach bug *Helicobacter pylori*, which causes ulcers. There are no such working vaccines for humans. (So far, the team has successfully tested their creation on mice.)

This is open-source biology, where intellectual property is shared freely. What's freely available to idealistic students, of course, would also be available to terrorists. Any number of blogs offer advice about everything from how to preserve proteins to the best methods for desalting DNA. Openness like that can be frightening, and there have been calls for tighter regulation—as well as suggestions that we stop this rampant progress before it becomes widely disseminated. Carlson, among many others, believes that strict regulations are unlikely to succeed. Several years ago, with very few tools but a working charge card, he opened his own biotechnology company, Biodesic, in the garage of his Seattle home—a biological version of the do-it-yourself movement that gave birth to so many computer companies, including Apple.

"It was literally in my garage," Carlson told me. The product enables the identification of proteins using DNA technology. "It's not complex, but I wanted to see what I could accomplish using mail order and synthesis." A great deal, it turned out. Carlson designed the molecule on his laptop, then sent the sequence to a company called Blue Heron that synthesizes DNA. Most instruments he needed could be purchased easily enough on eBay (or, occasionally, on LabX, a more specialized site for scientific equipment). "All you need is an Internet connection and a credit card," he said.

While nobody suggests that the field of synthetic biology should proceed without regulations, history has shown that they can produce consequences nobody really wants. "Strict regulation doesn't accomplish its goals," Carlson told me. "It's not an exact analogy, but look at Prohibition. What happened when government restricted the production and sale of alcohol? Crime rose dramatically. It became organized and powerful. Legitimate manufacturers could not sell alcohol, but it was easy to make in a garage—or a warehouse."

In 2002, the U.S. government began an intense effort to curtail the sale and production of methamphetamine. Before they did, the drug had been manufactured in many mom-and-pop labs throughout the country. Today it's mostly made on the black market; the laboratories have been centralized and the Drug Enforcement Administration says candidly that they know less about methamphetamine production than they did before. "The black market is getting blacker," Carlson said. "Crystal meth use is still rising, and all this despite restrictions." That doesn't mean strict control would ensure the same fate for synthetic biology. But it would be hard to see why it wouldn't.

The most promising technologies always present the biggest dangers. That's scary, but turning our backs on this opportunity would be scarier still. Many people suggest we do just that, though. Bill Joy, who founded Sun Microsystems, has frequently called for restrictions on the use of technology. "It is even possible that self-replication may be more fundamental than we thought, and hence harder—or even impossible—to control," he wrote in an essay for *Wired* magazine called "Why the Future Doesn't Need Us." "The only realistic alternative I see is relinquishment: to limit development of the technologies that are too dangerous by limiting our pursuit of certain kinds of knowledge."

Limit the pursuit of knowledge? When has that worked? Whom should we prevent from having information? And who would be the guardian of those new tools we consider too powerful to use? It would make more sense to do the opposite. Accelerate the development of technology and open it to more people and educate them to its purpose. Anything less would be Luddism. To follow Bill Joy's suggestion is to force a preventive lobotomy on the world. If Carlson is right—and I am sure that he is—the results would be simple to predict: power would flow directly into the hands of the people least likely to use it wisely, because fear and denialism are capable of producing no other result. This is a chance to embrace synthetic biology, and to end denialism.

To succeed we will have to stop conflating ideas and actions. There is no government conspiracy to kill American children with vaccines. I know that, and not because I believe blindly in our government or trust authority to tell me the truth. I don't. I know it because I believe in facts. Experts chosen to represent a specific point of view are cheerleaders, not scientists. And people who rely on them are denialists. No matter what happens on this planet—even if genetically engineered foods continue to feed us for centuries—there will be those who say the theoretical dangers outweigh the nourishment they can provide for billions of people. Impossible expectations are really just an excuse to seek comfort in lies. For all our fancy medical technology, Americans are no healthier and live no longer than citizens of countries that spend a fraction as much on health care. That can only change if alternatives are based on scientifically verifiable fact.

For synthetic biology to succeed we will also need an education system that encourages skepticism (and once again encourages the study of science). In 2008, students in Singapore, China, Japan, and Hong Kong (which was counted independently) all performed better on a standard international science exam, Trends in International Mathematics and Science Study, than American students. The U.S. scores have remained stagnant since 1995, the first year

the examination was administered. Adults are even less scientifically literate. Early in 2009, the results of a California Academy
of Sciences poll that was conducted throughout the nation revealed that only 53 percent of American adults know how long it
takes for the earth to revolve around the sun, and a slightly larger
number—59 percent—are aware that dinosaurs and humans
never lived at the same time.

Synthetic biologists will have to overcome this ignorance and
the denialism it breeds. To begin with, why not convene a new,
more comprehensive version of the Asilomar Conference, tailored
to the digital age and broadcast to all Americans? It wouldn't solve
every problem or answer every question—and we would need
many conversations, not one. But I can think of no better way for
President Obama to begin to return science to its rightful place in
our society. And he ought to lead that conversation through digital
town meetings that address both the prospects and perils of this
new discipline.

There would be no more effective way to vanquish denialism,
or help people adjust to a world that, as Drew Endy put it, is
surfing the exponential. It is not enough simply to tell people to
go back to school and learn about synthetic biology, or for that
matter, about how vaccines or vitamins or genomics work. Optimism only prevails when people are engaged and excited. Why
should we bother? Not to make *E. coli* smell like chewing gum or
fish glow in vibrant colors. Our planet is in danger, and the surest
way to solve the problem—and we can solve the problem—is to
teach nature how to do it.

The hydrocarbons we burn for fuel are really nothing more
than concentrated sunlight that has been collected by leaves and

trees. Organic matter rots, bacteria break it down, and it moves underground, where, after millions of years of pressure, it turns into oil and coal. At that point, we go dig it up—at huge expense and with disastrous environmental consequences. Across the globe, on land and sea, we sink wells and lay pipe to ferry our energy to giant refineries. That has been the industrial model of development, and it worked for nearly two centuries.

It won't work any longer, though, and we need to stop it.

The Industrial Age is in decline, eventually to be replaced by an era of biological engineering. That won't happen easily (or overnight), and it will never provide a magic solution to our problems. But what worked for artemisinin can work for many of the products we need in order to survive as a species. "We are going to start doing the same thing with bacteria that we do with pets," the genomic futurist Juan Enriquez said, describing our transition from a world that relied on machines to one that relies on biology. "A housepet is a domesticated parasite. . . . It has evolved to have an interaction with human beings. The same thing with corn"—a crop that didn't exist until we created it. "That same thing is going to start happening with energy. We are going to domesticate bacteria to process stuff inside a closed reactor to produce energy in a far more clean and efficient manner. This is just the beginning stage of being able to program life."

It is also the beginning of a new and genuinely natural environmental movement—one that doesn't fear what science can accomplish, but only what we might do to prevent it.

AFTERWORD

Shortly after I sent the manuscript of *Denialism* to my publisher in the spring of 2009, the Centers for Disease Control reported that two children in Southern California had developed a "febrile respiratory illness," caused by a flu virus—one that had never been seen in humans. The virus had also never been recognized in a pig; yet, in a colossal error of judgment, the CDC referred to the infection as a swine-flu virus. For millions of Americans those words stirred memories of one of the country's most notorious public health debacles: in 1976, army recruits at Fort Dix, New Jersey, became infected by a strain of influenza resembling the Spanish-flu virus of 1918, which killed more than fifty million people. The Ford administration, fearing the worst, attempted to vaccinate the entire nation, but the lethal epidemic never arrived. Among the millions who were vaccinated, however, a few suffered injury, and some even died—creating a legacy of fear, suspicion, and denial that have persisted to this day.

Something profound happened on the road between the 1976 and last year's pandemic. Americans are no longer so eager to follow orders from a man in a white lab coat. They have lost faith in authority, and who can blame them? History has shown pretty convincingly that skepticism and doubt serve us better than complacency or blind faith. Yet, facts still ought to matter when it comes to making critical decisions, and I wrote *Denialism* because too often facts are treated as if they are beside the point. The facts about this flu, for instance, were formidable: the CDC estimates as many as ninety million Americans were infected with the strain, called H1N1, that nearly four hundred thousand Americans were hospitalized and more than ten thousand died. A powerful and effective vaccine with an excellent safety profile was developed quickly, and millions were inoculated without incident. While any vaccine carries the possibility of an adverse reaction, the risks are minute compared to the dangers posed by the diseases those vaccines prevent. Yet, nearly half of American adults said that they had no intention of getting the HINI vaccine for themselves or for their children.

While that did not surprise me, the vehemence with which people rejected medical reality was breathtaking. Throughout the fall and early winter, angry callers flooded radio talk shows to denounce federal officials who supported universal vaccination. When people (and I am one of them) described vaccines as the most effective public health measure humans have ever created, the reaction was immediate and harsh. Reams of safety data, gathered carefully over years, didn't seem to convince people; nor did thousands of flu deaths. Official assurances (including those made by President Obama, who declared the possibility of a pandemic a national emergency) proved of similarly minor consequence. One can never predict the severity of an epidemic in advance, but when the H1N1 pandemic turned out to be relatively mild, instead of relief many Americans acted as if they had been cheated.

Denialism had gone mainstream.

Swine flu vaccine paranoia showed once again how frequently people embrace fear, intuition, emotion, and personal anecdote over scientific evidence backed by well-documented research. Much worse was soon to follow. Late in November, an English hacker stole and made public more than one thousand e-mails exchanged between scientists at the Climate Research Unit at the University of East Anglia. Climate-change skeptics rejoiced, claiming that the correspondence contained evidence that global warming was a fabrication. "Climategate" became the latest weapon in the denialist arsenal. A few of those e-mails were damning. They showed scientists willing to cut a corner; some were rude, dismissive, and arrogant. Scientists, like everyone else, can be that way. Yet, in the nearly sixty megabytes of stolen information, nobody could find a shred of evidence to challenge the mountain of facts that demonstrate, clearly and irrevocably, that the earth is warming rapidly or that human activity is the biggest reason why.

Again, facts were incidental. It was no surprise to see people like Mohammad Al Sabban using the e-mails to bolster his claim that "there is no relationship whatsoever between human activities and climate change." He is, after all, Saudi Arabia's lead climate negotiator, and the well-being of his country depends almost wholly on the continued destruction of fossil fuels. Al Sabban was hardly alone though. By early 2010 climate-change scientists were under attack on many fronts—and critics like Sarah Palin began to deploy the see-no-evil tactics that have for years been standard in the relentless war against scientific progress. Climate denialism started to look very much like the persistent, vehement arguments that smoking never caused lung cancer, or that chlorofluorocarbons didn't deplete the ozone layer, or that vaccines somehow caused autism; all groundless positions contradicted by years of evidence and experience.

To paraphrase Yogi Berra, climategate was déjà vu all over again. The issue was particularly significant to me because I had brushed only lightly over climate-change denialism in the book. And I was criticized by people who argued there was no more compelling or potentially damaging evidence of denialism in our time. I completely agree, of course; while we don't fully understand the mechanics of climate change, the fact that the earth is warming rapidly and dangerously has long been indisputable. And so, to my mind, is the cause: it's us. Humans. We've been choking the atmosphere with greenhouse gases for more than two centuries. The impact has been so obvious that it didn't seem to warrant a new round of debate.

For similar reasons I had paid little attention to AIDS denialism or to creationism; at this point, after millions of deaths, if people want to believe that HIV doesn't cause AIDS or that evolution is the magical concoction of a higher being, I can say little to change their minds. It seemed far more useful to focus on less obvious examples of denialism that also cause harm— like our obsession with useless (and often dangerous) dietary supplements and our growing conviction that organic produce contains exalted properties and, by itself, can feed the growing population of the world.

My mistake. I neglected the first law of denialism. The truth is NOT going to get in the way of people who are moved by faith, greed, fear, or desire to deny what they see. I should have known that. I've been watching and writing about this kind of behavior for years. It would be nice to chalk it all up to right-wing nuts with parochial economic interests, but that wouldn't be accurate.

And that brings us to the second truth of denialism: denialism transcends politics. Yes, opponents of evolution attack science, progress, and reality from the right. But the growing army of organic food fundamentalists, so eager to cast scientific data aside in their certainty that organic foods will save the world, hail from the other side of the political spectrum. The dietary supplement industry, propelled by the conviction that every American has the right to swallow any pill he or she can get his hands on, no matter how useless or damaging, represents the counter-cultural left and the libertarian right in equal measure. I wish I could argue that the most maddening denialists of all—those who see vaccines as threats to their children's health rather than bulwarks against terrible diseases—were poorly educated. They are not. I've had more arguments with Ivy League graduates about whether measles shots can cause autism (no connection has ever been demonstrated) or whether multiple vaccines can overpower an infant's immune system (they can't) than I care to recall. Sadly, the vaccine activists so willing to deny reality are some of the best-educated, most caring, thoughtful, and misguided people I have ever known.

This should have been a triumphant year in the battle against vaccine denialism: more than once, courts ruled that there is no demonstrable link between autism and vaccines. Scores of major studies have reached the same conclusion. Then, in February 2010, *The Lancet* retracted a study it had published more than a decade ago which was led by the now notorious British physician Andrew Wakefield. In the study, which was so seriously flawed that ten of its thirteen authors long ago retracted their contributions, Wakefield had connected the symptoms of autism directly to the measles-mumps-rubella vaccine.

That study launched a wave of fear and panic that continues despite the retraction of the study itself. Vaccine rates have plummeted, and measles—a disease that killed 160,000 people in the developing world last year—has returned to places that hadn't reported a case for years. The article fueled a massive, and remarkably vibrant, antivaccine movement in the United States, led by people who insist that there's a connection between autism and the inoculation. None of the studies carried out to test that thesis have found any such link; in fact, no correlation has ever been discovered. Children who have been vaccinated develop autism at the same rate as those who have not. But the damage Wakefield has done cannot be overstated, and vaccine denialism has become a central issue in American public health as a result.

The Lancet acted in February only because it no longer had a viable choice: a week earlier, Britain's General Medical Coun-cil issued a scathing statement detailing Wakefield's unethical methods. The council found him guilty of dishonesty and of "callous disregard" for the pain of the children in his study. By the end of May, Wakefield had—finally—been stripped of his British medical license.

So are we moving in the right direction? Perhaps, but the last time I walked by a bookstore, I couldn't help but notice a prom-inent new offering by Wakefield (with a foreword by the equally dangerous antivaccine partisan Jenny McCarthy). Ironically enough, he called it: *Callous Disregard: Autism and Vaccines— The Truth Behind a Tragedy.*

The real truth behind the tragedy is that denialism contin-ues to thrive.

ACKNOWLEDGMENTS

I have been lucky with editors throughout my career, but never more so than during my decade at the *New Yorker.* For much of that time I have worked with John Bennet, who, like all of the great ones, combines deft literary talent with unique psychiatric skills. I could not have made my way toward this book without his guidance. Among the many people at the magazine (past and present) who have also helped me I want to thank: Dorothy Wickenden, Henry Finder, Jeff Frank, Ann Goldstein, Elizabeth Pearson-Griffiths, Elizabeth Kolbert, Pam McCarthy, Sarah Larson, Julia Ioffe, Amelia Lester, Lauren Porcaro and Alexa Cassanos.

Several friends and colleagues read early portions of the book, and a few even waded through all of it. Naturally, my sins (and even my opinions) are not theirs; but the book would have been immeasurably weaker without them. In particular, I would like to thank Daniel Zalewski and Meghan O'Rourke. Somehow, they each found time to read chapters, and then provide detailed suggestions for how to make them better. My friend Richard Brody read the book as I wrote it, with great enthusiasm but a stern eye for faulty logic and broken sentences. Suzy Hansen served not only as an able fact checker, but also as a demanding reader. Any mistakes that survived her attention—or anyone else's—are my fault completely.

At the Penguin Press, Ann Godoff and Vanessa Mobley embraced the book from the moment I spoke to them about it. My editor, Eamon Dolan, joined the process after it began—but somehow seemed to understand what I was trying to do better than I did. He read thoughtfully, rapidly, and deeply. I don't know what more a writer could ask. I am also indebted to Nicole Hughes, Tracy Locke, Caroline Garner, and Leigh Butler, who doubles as a longtime friend.

Many friends have heard me drone on about this subject for years—encouraging me all the while. (And also arguing—which I tend to see as the same thing.) For their support and good cheer I would like to thank Gary Kalkut, Esther Fein, Gerry Krovatin, Sarah Lyall, Robert McCrum, Anne McNally, Richard Cohen, John Kalish, Jacob Weisberg, Deborah Needleman, Jacob Lewis, Sergey Brin, Anne Wojcicki and Alessandra Stanley.

When I was a child it often annoyed me that my parents, Howard and Eileen Specter, acted as if I could do anything. As I age, however, I have come to realize there is a role for blind devotion in this world—and I thank them for it profusely. My brother, Jeffrey, and his wife Yaelle, shouldered much family responsibility while I hid behind my laptop, and without that help there would have been no book.

Over the years, I have interviewed many people who have helped me form and refine the ideas in *Denialism*. It would be impossible to thank them all—and by trying I would only fail. Others, quite sensibly, preferred to remain anonymous. I was aided greatly early in the project by a lengthy discussion with Juan Enriquez—a man who knows denialism when he sees it and has rejected it with singular eloquence. I would also like to thank: Linda Avey, Esteban Gonzalez Burchard, Art Caplan, Rob Carlson, Joe Cerrell, George Church, June Cohen, John Elkington, Drew Endy, Ed Farmer, Tony Fauci, Jay Keasling, C. Everett Koop, Marie McCormick, Brian Naughton, Marion Nestle, Paul Offit, Neil Risch, Paul Saffo, Robert Shapiro, Eric Topol, Kari Stefansson and Eckard Wimmer.

Thousands of words of thanks have already been written on behalf of my friend and agent, Amanda Urban. They are insufficient. She combines intelligence, tenacity, vigor and loyalty – not to mention the occasional touch of fury—in a bundle unlike any other on the planet. Anna Quindlen has been my friend since newspapers used hot type—fortunately for me the friendship outlasted it. Anna, too, read early drafts and provided me with many notes—for which I am more grateful than she could know.

On my first day of work at the *Washington Post*, nearly twenty-five years ago, I noticed a tall man wandering aimlessly in the aisles, looking a bit like he belonged in another place. That place turned out to be the office of the editor of the *New Yorker*. David Remnick's leadership of the magazine has been praised by many others—and I can only add to the chorus. Nearly every Monday, proof of his gift turns up on newsstands and in hundreds of thousands of mailboxes. David is a remarkable editor—but an even better friend. Our conversation—often separated by continents—has lasted many years. He was the earliest and most consistent proponent of this book—and I thank him for that and for everything else. *Denialism* is dedicated to my daughter, Emma, who at the ripe old age of sixteen manages to teach me something new every day.

NOTES

Most of the information contained in this book comes either from interviews or from the large and constantly growing body of scientific research that addresses the subjects of each chapter. I will put footnotes on my Web site, www.michaelspecter.com.

I thought it might be useful here to point out at least some of the sources I found particularly compelling. Traditional journalists (a category that includes me) tend to deride blogs as so much unedited and contradictory noise. That's often true; but some of the most insightful science writing in America can be found on blogs these days—and I was lucky to have them at my disposal. Five in particular stand out as well-written, factually precise, and remarkably comprehensive: Aetiology, which focuses on evolu-

tion, epidemiology, and the implications of disease (http://sci
enceblogs.com/aetiology/); Respectful Insolence, a medical blog
that explains itself at the outset with the thoroughly accurate comment "A statement of fact cannot be insolent" (http://science
blogs.com/insolence/); Science-Based Medicine (http://www.sci
encebasedmedicine.org); Neurodiversity, which is almost certainly
the most complete archive of documents related to autism on the
Internet (http://www.neurodiversity.com/main.html); and Denialism (http://scienceblogs.com/denialism/). For some reason, I
didn't stumble upon the last of them until late in the process of
writing this book—but it's excellent.

1.
Vioxx and the Fear of Science

For a book that addresses the causes of our growing sense of disillusionment with
the American medical establishment, I would suggest John Abramson's *Overdosed America: The Broken Promise of American Medicine* (HarperPerennial,
rev. ed., 2008). There were two congressional hearings on Vioxx. Documents
pertaining to the first, held on November 19, 2004, by the Senate Finance
Committee, are available at http://finance.senate.gov/sitepages/hearing111804
.htm. Representative Henry A. Waxman convened hearings in the House on
May 5, 2005, which focused on how drugs are marketed in the United States.
All testimony and supporting material is available at http://waxman.house.gov/
News/DocumentSingle.aspx?DocumentID=122906. In retrospect, the initial
2001 study by Eric Topol and his colleagues, which appeared in the *Journal of
the American Medical Association*, stands out as restrained, well-reasoned, and
prescient. Unless you are an AMA member, though, you will have to buy it
(http://jama.ama-assn.org/cgi/content/full/ 286/22/2808). "What Have We
Learnt from Vioxx?" by Harlan M. Krumholz and several colleagues examines
the episode and its impact. The article, published by the *British Medical Journal*,
appeared in January 2007 (www.bmj.com/cgi/content/full/334/7585/120).
For two particularly useful discussions of eugenics, I would recommend Daniel J.
Kevles, *In the Name of Eugenics: Genetics and the Uses of Human Heredity* (Harvard

University Press, 1995). To get a sense of how a thoughtful scientist can follow reason and logic out the window (and take large segments of the world with him), there is no better place to go than to Francis Galton's *Hereditary Genius* (Prometheus Books, 1869).

2.
Vaccines and the Great Denial

For a deeply insightful primer on vaccines, the place to turn is Arthur Allen's *Vaccine: The Controversial Story of Medicine's Greatest Lifesaver* (Norton, 2007). Paul Offit not only invents vaccines, he writes about them with great authority. I am deeply indebted to his 2008 book *Autism's False Prophets: Bad Science, Risky Medicine, and the Search for a Cure* (Columbia University Press). The National Academy of Sciences, through the Institute of Medicine, has released two exhaustive reports on the safety of vaccines: *Immunization Safety Review: Measles-Mumps-Rubella Vaccine and Autism* (2001), and *Immunization Safety Review: Vaccines and Autism* (2004). Both are available through the NAS Web site (http://www .nationalacademies.org).

It would be hypocritical of me, in this book above all, to ignore those who reject the scientific consensus. Two places to begin: the National Vaccine Information Center (http://www.nvic.org), and David Kirby's book *Evidence of Harm: Mercury in Vaccines and the Autism Epidemic: A Medical Controversy* (St. Martin's Press, 2005). Kirby also maintains a robust collection of articles, testimony, and transcripts at http://www.evidenceofharm.com/index.htm.

3.
The Organic Fetish

The best book I have ever read about the ways in which genetically engineered and organic food relate to each other and to society is by the husband-and-wife team Pamela Ronald and Raoul Adamchak, *Tomorrow's Table: Organic Farming, Genetics, and the Future of Food* (Oxford University Press, 2008). Adamchak is an organic farmer and Ronald a plant geneticist. Their knowledge, sophistication, and priorities ought to provide at least some evidence that seemingly irreconcilable differences are not impossible to resolve. (Ronald also maintains a fascinating blog by the same name, http://pamelaronald.blogspot.com.)

Everything Marion Nestle writes is worth reading (usually more than once). I particularly recommend *Food Politics: How the Food Industry Influences Nutrition and Health* (University of California Press, 2002) and *What to Eat* (North Point Press, 2006). Denise Caruso runs the Hybrid Vigor Institute. Her call to excess caution seems unwarranted to me, but nobody makes the argument better or more thoroughly: *Intervention: Confronting the Real Risks of Genetic Engineering and Life on a Biotech Planet* (Hybrid Vigor Press, 2006).

For data on agricultural production, hunger, or development in Africa, I suggest that any interested reader look at the World Bank's *2008 World Development Report: Agriculture for Development.* (The URL for this report is almost comically long. It would be far easier to go to the bank's general site, www.worldbank .org, and type "2008 world development report" into the search box.) Among the other studies I have found useful: the Pew Charitable Trust 2008 report *Putting Meat on the Table: Industrial Farm Animal Production in America* (http:// www.ncifap.org/) and the Rockefeller Foundation's 2006 study *Africa's Turn: A New Green Revolution for the 21st Century* (www.rockfound.org/library/africas_ turn.pdf). The annual report of the Food and Agriculture Organization of the United Nations always addresses these issues, but never more directly than the 2004 study *Agricultural Biotechnology: Meeting the Needs of the Poor?* (www .fao.org/es/esa/pdf/sofa_flyer_04_en.pdf). Finally, Louise O. Fresco has written often and revealingly about issues of food security in the developing world. See particularly her report, last updated in 2007, *Biomass, Food & Sustainability: Is There a Dilemma?* (www.rabobank.com/content/images/Biomass_food_and_ sustainability_tcm43-38549.pdf).

There are many discussions of the "precautionary principle," fear, and the idea of risk. Four stand out to me: Cass Sunstein's *Laws of Fear: Beyond the Precautionary Principle* (Cambridge University Press, 2005); Lars Svendsen's *A Philosophy of Fear* (Reaktion Books, 2008); Peter L. Bernstein's *Against the Gods: The Remarkable Story of Risk* (Wiley, 1996); and Leonard Mlodinow's *The Drunkard's Walk: How Randomness Rules Our Lives* (Pantheon, 2008).

4.
The Era of Echinacea

The Cochrane Collaboration (www.cochrane.org), through its Database of System-atic Reviews, comes as close as possible to providing authoritative information in a field that needs it badly. In addition, the National Center for Complemen-tary Medicine, the Harvard School of Public Health, and the Memorial Sloan-Kettering Cancer Center each offer information on vitamins and supple-

ments at http://nccam.nih.gov, http://www.hsph.harvard.edu, and http://www
.mskcc.org/mskcc/html/1979.cfm respectively, as of course do many other
institutions.

The two best recent treatments of alternative health have both been written or edited
by Ernst Edzard, who is professor of complementary medicine at the universities
of Exeter and Plymouth. The first, written with Simon Singh, is *Trick or Treatment: The Undeniable Facts about Alternative Medicine* (Norton, 2008). Edzard
also edited *Healing, Hype or Harm? A Critical Analysis of Complementary or Alternative Medicine* (Societas, 2008). For the other side of the story, Andrew Weil
is the man to see. He is prolific, but one might begin with *Healthy Aging: A
Lifelong Guide to Your Physical and Spiritual Well-Being* (Knopf, 2005).

For a disciplined and opinion-free history of vitamin regulation in America, see the
1988 *Surgeon General's Report on Nutrition and Health.* The managing editor was
Marion Nestle, and the 750-page report is available at her Web site, among other
places (www.foodpolitics.com/wp-content/uploads/surgeon-general.pdf).

I try to remain open-minded on all scientific issues, but there are limits. Those
eager to explore the phenomenon of AIDS denialism are on their own. Anyone seeking to understand the actual roots of the disease, or its natural progression, however, can start at www.aidstruth.org—which lives up to its name.

5.

Race and the Language of Life

For a general argument on the issue of race and ethnic background in medical treat-
ment, there is the 2003 piece by Burchard and Risch et al., "The Importance of
Race and Ethnic Background in Biomedical Research and Clinical Practice." For
an abstract and an extensive list of subsequent papers on the topic go to http://
content.nejm.org/cgi/content/short/348/12/1170; Sandra Soo-Jin Lee's essay
"Racializing Drug Design: Implications of Pharmacogenomics for Health
Disparities," in the December 2005 issue of the *American Journal of Public
Health,* is a smart discussion of race and genomics (www.ajph.org/cgi/reprint/
AJPH.2005 .068676v2.pdf). The New York University sociologist Troy Duster
has written widely on the topic as well; see *Backdoor to Eugenics* (Routledge,
2003), among many other publications. Robert S. Schwartz argues that geno-
mics has turned the concept of race into a dangerous anachronism in his "Racial
Profiling in Medical Research," *New England Journal of Medicine* 344, no. 18
(2001). It can be purchased at the journal's Web site (http://content.nejm.org).
The best short explanatory book I have ever read on the subject of genetics is Adrian

Woolfson's *An Intelligent Person's Guide to Genetics* (Overlook Press, 2004). Two
other books have proven valuable to me: James Schwartz, *In Pursuit of the Gene:
From Darwin to DNA* (Harvard University Press, 2008), and Barry Barnes
and John Dupré, *Genomes and What to Make of Them* (University of Chicago
Press, 2008).

6.
Surfing the Exponential

As I note in the book, the phrase "surfing the exponential" comes from Drew Endy
of Stanford University. The best study on the topic is *New Life, Old Bottles:
Regulating First-Generation Products of Synthetic Biology* by Michael Rodemeyer, a
former director of the Pew Charitable Trust's Initiative on Food and Biotechnology. This report, issued in March 2009 under the auspices of the Woodrow
Wilson International Center for Scholars, can be obtained from the Synthetic
Biology Project (http://www.synbioproject.org/library/publications/archive/
synbio2/).

The ETC Group (Action Group on Erosion, Technology and Concentration) has
taken the lead in calling for stricter oversight of this new discipline. The group
poses thoughtful questions that demand thoughtful answers. On December 8,
2008, Steward Brand's Long Now Foundation sponsored an unusually amicable
debate between ETC's Jim Thomas and Endy. The conversation provides a
thorough airing of the issues and can be purchased on DVD at Amazon.com
(the podcast is also available at no charge: http://fora.tv/media/rss/Long_Now_
Podcasts/podcast-2008-11-17-synth-bio-debate.mp3).

ETC has released many studies, all of which can be found on the group's homepage
(http://www.etcgroup.org/en/issues/synthetic_biology.html). The most important and comprehensive of them, *Extreme Genetic Engineering*, is here (http://
www.etcgroup.org/en/issues/synthetic_biology.html).

Scientists are often accused of ignoring the ethical implications of their work. It is
worth nothing, then, that Craig Venter—the genomic world's brashest brand
name—embarked on a yearlong study of the ethical and scientific issues in synthetic biology before stepping into the lab. *Synthetic Genomics: Options of Gov-
ernance,* by Michele S. Garfinkel, Drew Endy, Gerald L. Epstein, and Robert M.
Friedman, is available at www.jcvi.org/cms/research/projects/syngen-options/
overview/, and the technical reports that were commissioned for the study can
be found at http://dspace.mit.edu/handle/1721.1/39658.

The scientific roots of synthetic biology are explored in Philip J. Pauly's book
Controlling Life: Jacques Loeb and the Engineering Ideal in Biology (Oxford University Press, 1987). It's expensive and hard to find; but it is out there. I would

also recommend Michael Rogers's book about the early days of recombinant DNA technology, *Biohazard* (Knopf, 1979). That, too, is difficult to find. For anyone inclined to wonder why Eckard Wimmer created a synthetic polio virus, I suggest reading his 2006 article on the implications of the research, published in the *Journal of the European Molecular Biology Organization*, "The Test-Tube Synthesis of a Chemical Called Poliovirus." A free, full-text version of the article can be found at http://www.pubmedcentral.nih.gov/articlerender.fcgi?tool=pubmed&pubmedid=16819446.

BIBLIOGRAPHY

Abramson, John, MD. *Overdosed America: The Broken Promise of American Medicine*. New York: HarperCollins, 2008.

Allen, Arthur. *Vaccine: The Controversial Story of Medicine's Greatest Lifesaver*. New York: W. W. Norton, 2007.

Bacon, Francis. *The New Organon; or, True Directions Concerning the Interpretation of Nature*. London, 1620.

Barnes, Barry, and John Dupré. *Genomes and What to Make of Them*. Chicago: University of Chicago Press, 2008.

Basalla, George. *The Evolution of Technology*. Cambridge, UK: Cambridge University Press, 1988.

Bausell, R. Barker. *Snake Oil Science: The Truth about Complementary and Alternative Medicine*. Oxford, UK: Oxford University Press, 2007.

Carlson, Rob. *Biology Is Technology: The Promise, Peril, and Business of Engineering Life*. Cambridge, MA: Harvard University Press, 2009.

Caruso, Denise. *Intervention: Confronting the Real Risks of Genetic Engineering and Life on a Biotech Planet.* San Francisco: Hybrid Vigor Press, 2006.

Chesterton, G. K. *Eugenics and Other Evils.* Seattle: Inkling Books, 2000.

Culshaw, Rebecca. *Science Sold Out: Does HIV Really Cause AIDS?* Berkeley, CA: North Atlantic Books, 2007.

Dawkins, Richard. *The Selfish Gene.* Oxford, UK: Oxford University Press, 1989.

Darwin, Charles. *On the Origin of Species.* London: John Murray, 1859.

Duster, Troy. *Backdoor to Eugenics.* New York: Routledge, 2003.

Ehrlich, Paul R. *The Population Bomb.* New York: Ballantine Books, 1968.

———, and Anne H. Ehrlich. *The Dominant Animal: Human Evolution and the Environment.* Washington, DC: Island Press, 2008.

Fukuyama, Francis. *Our Posthuman Future: Consequences of the Biotechnology Revolution.* New York: Picador, 2002.

Galton, Francis. *Hereditary Genius.* Amherst, MA: Prometheus Books, 1869.

Gessen, Masha. *Blood Matters: From Inherited Illness to Designer Babies, How the World and I Found Ourselves in the Future of the Gene.* Orlando, FL: Houghton Mifflin Harcourt, 2008.

Goldstein, David B. *Jacob's Legacy: A Genetic View of Jewish History.* New Haven, CT: Yale University Press, 2008.

Goodman, Alan H., Deborah Heath, and M. Susan Lindee. *Genetic Nature/Culture: Anthropology and Science Beyond the Two-Culture Divide.* Berkeley: University of California Press, 2003.

Grace, Eric S. *Biotechnology Unzipped: Promises and Realities.* Washington, DC: Joseph Henry Press, 2006.

Graham, Loren R. *The Ghost of the Executed Engineer: Technology and the Fall of the Soviet Union.* Cambridge, MA: Harvard University Press, 1996.

Ham, Ken, and Charles A. Ware. *Darwin's Plantation: Evolution's Racist Roots.* Green Forest, AR: Master Books, 2007.

Harford, Tim. *The Logic of Life: The Rational Economics of an Irrational World.* New York: Random House, 2008.

Hind, Dan. *The Threat to Reason.* New York: Verso, 2007.

Holloway, David. *Stalin and the Bomb: The Soviet Union and Atomic Energy, 1939–1956.* New Haven, CT: Yale University Press, 1994.

Hope, Janet. *BioBazaar: The Open Source Revolution and Biotechnology.* Cambridge, MA: Harvard University Press, 2008.

Hughes, Thomas P. *American Genesis: A Century of Invention and Technological Enthusiasm, 1870–1970.* New York: Penguin, 1990.

———. *Human-Built World: How to Think about Technology and Culture.* Chicago: University of Chicago Press, 2004.

Jones, Steve. *The Language of Genes.* New York: Anchor, 1995.

Kevles, Daniel J. *In the Name of Eugenics: Genetics and the Uses of Human Heredity.* Cambridge, MA: Harvard University Press, 1995.

Kirby, David. *Evidence of Harm: Mercury in Vaccines and the Autism Epidemic; A Medical Controversy.* New York: St. Martin's Press, 2005.

Lebo, Lauri. *The Devil in Dover: An Insider's Story of Dogma v. Darwin in Small-Town America.* New York: New Press, 2008.

Levin, Yuval. *Imagining the Future: Science and American Democracy.* New York: Encounter Books, 2008.

Levy, David. *Love + Sex with Robots: The Evolution of Human-Robot Relationships.* New York: HarperCollins, 2007.

Loomis, William F. *Life as It Is: Biology for the Public Sphere.* Berkeley: University of California Press, 2008.

McKibben, Bill. *The End of Nature.* New York: Random House, 1989.

McNeill, William H. *Plagues and Peoples.* New York: Anchor, 1977.

McWilliams, James E. *American Pests: The Losing War on Insects from Colonial Times to DDT.* New York: Columbia University Press, 2008.

Menzies, Gavin. *1421: The Year China Discovered America.* New York: HarperCollins, 2002.

Miller, Henry I., and Gregory Conko. *The Frankenfood Myth: How Protest and Politics Threaten the Biotech Revolution.* Westport, CT: Praeger, 2004.

Mlodinow, Leonard. *The Drunkard's Walk: How Randomness Rules Our Lives.* New York: Pantheon, 2008.

Nestle, Marion. *Food Politics. How the Food Industry Influences Nutrition and Health.* Berkeley: University of California Press, 2002.

————. *What to Eat*. New York: North Point, 2006.

Noble, David F. *Progress Without People: In Defense of Luddism*. Chicago: Charles H. Kerr, 1993.

Null, Gary. *AIDS: A Second Opinion*. New York: Seven Stories Press, 2002.

Offit, Paul A. *The Cutter Incident: How America's First Polio Vaccine Led to the Growing Vaccine Crisis*. New Haven, CT: Yale University Press, 2005.

————. *Vaccinated: One Man's Quest to Defeat the World's Deadliest Diseases*. New York: HarperCollins, 2007.

————. *Autism's False Prophets: Bad Science, Risky Medicine, and the Search for a Cure*. New York: Columbia University Press, 2008.

Paarlberg, Robert. *Starved for Science: How Biotechnology Is Being Kept Out of Africa*. Cambridge, MA: Harvard University Press, 2008.

Pauly, Philip J. *Controlling Life: Jacques Loeb and the Engineering Ideal in Biology*. New York: Oxford University Press, 1987.

Pennock, Robert T., ed. *Intelligent Design Creationism and Its Critics: Philosophical, Theological, and Scientific Perspectives*. Cambridge, MA: MIT Press, 2001.

Perrin, Noel. *Giving Up the Gun: Japan's Reversion to the Sword, 1543–1879*. Boston: David. R. Godine, 1979.

Pinker, Steven. *The Stuff of Thought: Language as a Window into Human Nature*. New York: Viking Penguin, 2007.

Pollan, Michael. *The Omnivore's Dilemma: A Natural History of Four Meals*. New York: Penguin Press, 2006.

————. *In Defense of Food: An Eater's Manifesto*. New York: Penguin Press, 2008.

Rees, Martin. *Our Final Century: Will Civilisation Survive the Twenty-first Century?* London: Arrow Books, 2004.

Regis, Ed. *What Is Life? Investigating the Nature of Life in the Age of Synthetic Biology*. New York: Farrar, Straus and Giroux, 2008.

Ronald, Pamela C., and Raoul W. Adamchak. *Tomorrow's Table: Organic Farming, Genetics, and the Future of Food*. Oxford, UK: Oxford University Press, 2008.

Rogers, Michael. *Biohazard*. New York: Knopf, 1997.

Rose, Nikolas. *The Politics of Life Itself: Biomedicine, Power, and Subjectivity in the Twenty-first Century*. Princeton, NJ: Princeton University Press, 2007.

Sarich, Vincent, and Frank Miele. *Race: The Reality of Human Differences*. Boulder, CO: Westview Press, 2004.

Satel, Sally. *PC, M.D.* New York: Basic Books, 2000.

Schwartz, James. *In Pursuit of the Gene: From Darwin to DNA*. Cambridge, MA: Harvard University Press, 2008.

Singh, Simon, and Edzard Ernst. *Trick or Treatment: The Undeniable Facts about Alternative Medicine*. New York: W. W. Norton, 2008.

Smith, Jeffrey M. *Genetic Roulette: The Documented Health Risks of Genetically Engineered Foods*. Fairfield, CT: Yes! Books, 2007.

Smith, Merritt Roe, and Leo Marx, eds. *Does Technology Drive History? The Dilemma of Technological Determinism*. Cambridge, MA: MIT Press, 1994.

Speth, James Gustave. *The Bridge at the Edge of the World: Capitalism, the Environment, and Crossing from Crisis to Sustainability*. New Haven, CT: Yale University Press, 2008.

Starr, Paul. *The Social Transformation of American Medicine*. New York: Basic Books, 1984.

Sunstein, Cass R. *Laws of Fear*. Cambridge, UK: Cambridge University Press, 2005.

Svendsen, Lars. *A Philosophy of Fear*. London: Reaktion Books, 2008.

Venter, J. Craig. *A Life Decoded: My Genome: My Life*. New York: Viking, 2007.

Walker, Gabriele, and Sir David King. *The Hot Topic: What We Can Do about Global Warming*. Orlando, FL: Houghton Mifflin Harcourt, 2008.

Weil, Andrew. *Spontaneous Healing: How to Discover and Embrace Your Body's Natural Ability to Heal Itself*. New York: Ballantine Books, 2000.

———. *Healthy Aging: A Lifelong Guide to Your Physical and Spiritual Well-Being*. New York: Knopf, 2005.

Weisman, Alan. *The World Without Us*. New York: St. Martin's Press, 2007.

Wilmut, Ian, and Roger Highfield. *After Dolly: The Uses and Misuses of Human Cloning*. New York: W. W. Norton, 2006.

Wilson, Edward O. *The Future of Life*. New York: Knopf, 2002.

Woolfson, Adrian. *An Intelligent Person's Guide to Genetics*. London: Duckworth/ Overlook, 2004.

INDEX